SAN FRANCISCO
PUBLIC LIBRARY

SCIENCE & TECHNOLOGY
DEPARTMENT

REFERENCE BOOK

Not to be taken from the Library

SCIENCE IN PROGRESS
TWELFTH SERIES

THE SOCIETY OF THE SIGMA XI
AND THE
SCIENTIFIC RESEARCH SOCIETY OF AMERICA

DEVOTED TO THE
ENCOURAGEMENT OF RESEARCH IN SCIENCE

SIGMA XI – RESA
National Lectureships
1959 and 1960

SCIENCE IN PROGRESS
TWELFTH SERIES

TWELFTH SERIES

Edited by WALLACE R. BRODE

Science in Progress

BY

GEORGE C. KENNEDY
WILLEM J. LUYTEN
JOHN VERHOOGEN
PAUL DELAHAY
HAROLD G. CASSIDY
R. F. DAWSON
J. HERBERT TAYLOR
EMIL WITSCHI
ROBERT C. ELDERFIELD
RALPH H. WETMORE
HARRY F. HARLOW
BARNETT F. DODGE

New Haven and London, 1962
YALE UNIVERSITY PRESS

Copyright © 1962 by Yale University.
Set in Baskerville type and
printed in the United States of America by
The Carl Purington Rollins Printing Office
of the Yale University Press.
All rights reserved. This book may not be
reproduced, in whole or in part, in any form
(except by reviewers for the public press),
without written permission from the publishers.
Library of Congress catalog card number: 39–14778
Published with assistance from the Louis Stern Memorial Fund.

CONTENTS

Preface   WALLACE R. BRODE   vii

1. The Origin of Continents, Mountain Ranges, and Ocean Basins
   GEORGE C. KENNEDY   1

2. White Dwarfs and Stellar Evolution
   WILLEM J. LUYTEN   23

3. Temperatures Within the Earth
   JOHN VERHOOGEN   39

4. Reflections on the Cultivation of Science
   PAUL DELAHAY   77

5. The Problem of the Sciences and the Humanities. A Diagnosis and a Prescription
   HAROLD G. CASSIDY   93

6. Biosynthesis of the Nicotiana Alkaloids
   R. F. DAWSON   117

7. Chromosome Reproduction and the Problem of Coding and Transmitting the Genetic Heritage
   J. HERBERT TAYLOR   145

8. Sex Reversal in Animals and in Man
   EMIL WITSCHI   171

9. Australian Trees and High Blood Pressure
   ROBERT C. ELDERFIELD   195

10. Morphogenesis in Plants—A New Approach
    RALPH H. WETMORE   217

11. The Development of Learning in the Rhesus Monkey
    HARRY F. HARLOW   239

12. Fresh Water from Saline Waters. An Engineering Research Problem
    BARNETT F. DODGE   271

Index   327

PREFACE

THE TWELFTH VOLUME of the series *Science in Progress* includes the Sigma Xi–RESA National Lectures given in 1959 and 1960. When taken together with the previous volumes they present a quarter century of national lectures, sponsored first by the Society of the Sigma Xi and subsequently jointly sponsored by Sigma Xi and the Scientific Research Society of America. In this 25-year period there have been some 130 lectures published in the 12 volumes under the editorship successively of George A. Baitsell, Hugh Taylor, and Wallace R. Brode. The collection of the Sigma Xi–RESA National Lectures is nearly complete, for only a few, usually for very good reasons, are not prepared for publication in the *American Scientist* and *Science in Progress*. The collected lectures provide an important reference series for the scholar, student, and research worker.

The general policy of the Committee on National Lectureships is to select experts in important and timely areas of science and to advise them that their presentation should not be directed primarily toward specialists in their own area, but rather to inform other scientists and the technically understanding public of the advances in the lecturer's field.

The popularity of the national lectures is well attested by the increasing requests from Sigma Xi chapters and clubs and RESA branches to serve as sponsors of the lectures in their communities. The lecturer's endurance and available time usually limit a tour to 20 to 25 lectures and hence, with some 300 chapters, clubs, and branches, it has been found necessary to increase the number of lectureships from the original four to eight, and more may be added if the demand continues to grow.

The Executive Committee of Sigma Xi has taken cognizance of this recent increase by authorizing a proportionately stepped-up production rate of the *Science in Progress* series. The future issuance on essentially an annual instead of biennial basis will to a considerable degree remove the feeling of some of the contributors that the series may not, in a rapidly moving field, be keeping up with the latest developments. Some authors indicate in their written manuscripts, and even in inserts in proof, that important changes have taken place since their lectures were given which may alter views expressed in the lectures. To a considerable degree these papers are archival, in that they represent the state of the art at the time of the lecture. It is hoped that the accelerated rate of publication of future volumes will do much to improve the timely status of the contributions.

The current volume reflects the continuing interest in terrestrial problems that first received attention during the International Geophysical Year. The problems of the scientist's relations with his fellow human beings have brought forth valuable reflections by scientists who were generally expected to be so wrapped up in their technology that they would have little time to spare for the larger problems of the welfare of our civilization. The discussions in this volume on animal behavior, chromosomes, and organic synthesis point to a growing awareness of the importance of the life sciences and the cooperative fields of biochemistry, biophysics, and psychology.

BIOGRAPHICAL DATA

HAROLD G. CASSIDY, Ph.D., Yale (1939). Professor of chemistry at Yale University. A polymer chemist whose principal publications have been in the areas of chromatographic adsorption and the preparation and study of electron exchange polymers. He has developed a new and important facet of science in his studies of cold war problems and the relations between science and the humanities.

R. F. DAWSON, Ph.D., Yale (1938). John Torrey Professor of botany at Columbia University. A tobacco plant expert who has spent considerable time in the study of tropical plants in Central America. His researches include studies of the biosynthesis of plant alkaloids, chemistry and physiology of the tobacco plant, and tropical drug plants.

PAUL DELAHAY, Ph.D., Oregon (1948). Boyd Professor of chemistry at Louisiana State University. His research work has been concentrated on electrochemical processes, diffusion, corrosion, high-temperature metal-gas reactions, and new instrumental analysis methods. With many of our leading scientists, he has chosen to delve into the cultural aspects of science.

BARNETT F. DODGE, Sc.D., Harvard (1925). Professor of chemical engineering at Yale University since 1925. Past president of the American Institute of Chemical Engineers. An authority on thermodynamic properties at high pressures, rate of heat transfer, and treatment of saline water and industrial waste water.

ROBERT C. ELDERFIELD, Ph.D., Massachusetts Institute of Technology (1930). Professor of Chemistry at the University of Michigan. Member of the National Academy of Sciences and chairman of the Division of Chemistry of the National Research Council. An authority in the field of heterocyclic organic compounds, chemotherapy, and the structure and synthesis of cardiac drugs.

HARRY F. HARLOW, Ph.D., Stanford (1930). Professor of psychology at the University of Wisconsin since 1930. Member of the National Academy of Sciences and past chairman of the Division of Anthropology and Psychology of the National Research Council. Editor of the *Journal of*

Comparative and Physiological Psychology and past president of the American Psychological Association. An authority on primate learning and behavior, long-term effects of radiation, and development of affection.

GEORGE C. KENNEDY, Ph.D., Harvard (1946). Professor of geophysics at the University of California at Los Angeles. Served with the U. S. Geological Survey in Alaska. His research fields include volcanology, solubility in gas phase, phase systems involving silicates, and the physics of high pressures.

WILLEM J. LUYTEN, Ph.D., Leiden (1921). Professor of astronomy at the University of Minnesota since 1931. An internationally known astronomer with special interest in solar eclipses and stellar motions. His research deals with Zwicky stars, white dwarfs, and the origin of the solar system.

J. HERBERT TAYLOR, Ph.D., Virginia (1944). Professor of cell biology at Columbia University. Recognized for his studies on chromosome duplication involving radioactive markers. His research also includes work in cytogenetics, physiological relations of diploid and polyploid plants, and cytotaxonomy of the family Oleaceae.

JOHN VERHOOGEN, Ph.D., Stanford (1936). Professor of geology at the University of California at Berkeley, with a distinguished background of field experience in the Belgian Congo. Member of the National Academy of Sciences and authority on the earth's magnetic field, volcanology, geophysics of solid earth, and the properties of rocks which give rise to the earth's temperatures.

RALPH H. WETMORE, Ph.D., Harvard (1924). Retiring as emeritus professor of botany in 1962 after some 40 years at Harvard. Nationally and internationally recognized with many honors, including membership in the National Academy of Sciences and the presidency of the Botanical

Society of America. He is an expert in plant morphology, development of vascular plants, and the problems of plant anatomy.

EMIL WITSCHI, Ph.D., Munich (1913). Emeritus professor of zoology at the University of Iowa, where he taught for more than 30 years. Past president of the American Society of Zoologists. A world-recognized and honored authority in the fields of genetics, embryology, and endocrinology.

The Editor wishes to express his appreciation to Hugh Taylor, editor of the *American Scientist* and former editor of the *Science in Progress*, for his helpful advice and suggestions. Recognition should also be given, in the conclusion of the first quarter century of this series, to the effective editorial leadership of George A. Baitsell, who established the series and edited the first nine volumes.

Acknowledgment is also due to the authors of the separate lectures and to the staff of the Yale University Press for assistance in the preparation of this volume. The authors have quoted freely from their own previously published works and have referred to other sources in technical journals. Prior approval was sought and acknowledged at the time of publication in the *American Scientist* and is again made to the many sources drawn on and, to the Editor's best knowledge, properly credited by the separate authors.

WALLACE R. BRODE

Washington, D.C.
February 1962

THE ORIGIN OF CONTINENTS, MOUNTAIN RANGES, AND OCEAN BASINS*

By GEORGE C. KENNEDY
University of California
at Los Angeles

ONE OF THE unexpected discoveries in earth science in the previous century was that of a fundamental difference between continents and ocean basins. Ocean basins are not merely the low-lying parts of the earth's surface flooded by salt water but are great, relatively steep-sided, structural depressions. In fact, there is too much water for the size of the ocean basins, and parts of the edges of the continents are now flooded and probably have been flooded through a great deal of the geologic past. A typical continental mass with adjacent ocean basins is shown in Figure 1A.

Precise measurements of gravitational attraction in major mountain ranges, continental areas, and over the ocean basins showed an even more unexpected feature. The continents and mountain ranges do not represent extra loads of rock superimposed upon the earth's crust, but are masses of lighter rock floating in a denser substrate. An iceberg floats

* Publication 122, Institute of Geophysics, University of California, Los Angeles, Calif.

above the water in much the same fashion, buoyed up by deep submerged roots. The great mountain ranges of the world and the continental masses similarly have deep roots of light rock penetrating into the denser crust, and thus the

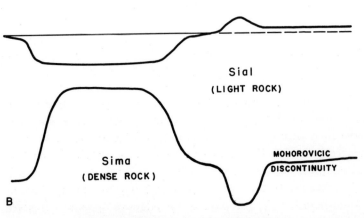

Fig. 1. Profiles through the earth's crust.

mountain ranges and continents float at elevations appropriate to the depth and size of these submerged light roots. Thus, all the major features of relief of the surface of the earth show mirror image features within the crust, much as is indicated in Figure 1B. The major mountain ranges float at high elevations because they are buoyed up by light rocks. The continents float at intermediate elevations with roots of intermediate depth, and the deep oceans are underlain by thin layers of light rock.

Seismologists, from the study of earthquake waves, have shown that the earth's mantle is solid to the depth of the outer core, some 2,900 kilometers. The observation that large mountain ranges and continental masses float on the crust of the earth at elevations appropriate to the size and density of their roots implies that rocks at shallow depths in the earth's mantle, although solid, have little strength and can flow in response to small stresses, given sufficient time. This deduction is strengthened by the observation that rocks in deep, eroded, old mountain chains are intensely contorted and folded, plain evidence that, at high pressures, solid rocks can flow readily and do not have great strength.

Recognition that continental rocks are lighter and more buoyant than oceanic rocks gave rise to the concept that the crust of the earth is made of two contrasting materials: sialic material, rich in silicon, the alkalies, and aluminum, making up the continents; and simatric material, richer in iron and magnesium, making up the denser rocks below the floor of the ocean and lying under the sial of the continents. The sial is assumed to be granite or granodiorite in composition, and the sima is assumed to be basaltic in composition.

Early in this century the Yugoslavic seismologist, Mohorovičić, obtained evidence from seismograms that earthquake waves, traveling a few tens of kilometers below the surface of the earth, gave records showing sharply higher speeds for both shear and compressional waves than do earthquake waves traveling near the surface. This indicated an abrupt change in rock types at a few tens of kilometers under the continents and at a few kilometers under the oceans. In recent years extensive studies have produced a fairly clear general picture of the nature and depth of this level of change, or discontinuity, under the continents, mountain ranges, and ocean basins. This discontinuity, called the Mohorovičić or M discontinuity, is at a general depth of 30 to 40 kilometers under the continents, but may be as deep as 60 kilometers under the roots of major mountain chains. It is as shallow as 4 to 5 kilometers below the floor of the deeper parts of the ocean. The discovery of the M discontinuity seemed to confirm the notion that the crust is fundamentally made up of two different kinds of rock material. The discontinuity itself appears to be the boundary between these, the sialic rocks above and the simatic rocks below. The rocks below the M discontinuity have seismic velocities and densities which suggest that they may be even richer in magnesium and iron than normal sima of basaltic composition. Consequently they are, by some, called ultrasima. However, throughout this paper the word sial will be applied to the lower-velocity rocks above the M discontinuity, including the range of basalts to granites, and the word sima will be applied to the denser rocks below the M discontinuity.

Prior to and along this general picture, the concept of isostasy developed. This is the notion, previously discussed, that the lighter continental rocks float at an appropriate depth, depending on their mass and mean density, in a denser substratum. As rock is eroded from the tops of continents and mountain ranges they tend to float up higher and higher, renewing their relief, permitting erosion to continue.

Four facts, however, sharply contradict this picture of a sialic crust of varying thickness floating on a simatic substratum of different chemical composition and different density:

1. Large areas of continents, long near sea level, have been uplifted many thousands of feet into the air. Further, this uplift seems to have taken place rather rapidly in terms of geologic time.

2. Sediments of low density, filling troughs along the margins of continents, apparently are able to subside into this higher density substratum.

3. Inasmuch as radioactive, heat-producing elements are

associated with sialic rocks, one might expect heat flow through the thicker parts of the earth's crust to be much greater than through the thinner parts. However, as a first approximation, heat flow through the crust of the earth is approximately the same through continents, mountain ranges, and ocean basins.

4. The lifetime of continents and mountain ranges is vastly greater than rates of erosion would suggest.

Let us examine each of these apparent facts and their consequences on the hypothesis of sialic continents floating on a simatic basin. The problem of the uplift of large plateau areas is one that has puzzled students of the earth's crust for a very long time. Regions at sea level or near sea level may, over a relatively short geologic time span such as a few million years, be uplifted several thousands of feet. The Colorado plateau and adjacent highlands is an example. Here, an area of approximately 250,000 square miles, that apparently stood at sea level for several hundreds of millions of years, was uplifted approximately one mile vertically some 40,000,000 years ago in early mid-Tertiary time and is still a high plateau. The Grand Canyon of the Colorado has been carved through this great uplifted plateau.

Given an earth with sialic continents floating in a denser simatic substratum, what mechanism would cause a large volume of low-standing continent to rise rapidly a mile into the air? Furthermore, evidence from gravity surveys suggest that the rocks underlying the Colorado plateau are in isostatic balance; that is, this large area is floating at its correct elevation in view of its mass and density. Recent seismic evidence confirms this, in that the depth to the M discontinuity under the Colorado plateau is approximately 10 kilometers greater than over most of continental North America. Thus, appropriate roots of light rock extend into the dense substratum to account for the higher elevation of the Colorado plateau. We have then a double-ended mystery, for the Colorado plateau seems to have grown downward at the same time that its emerged part rose upward. This is just as startling as it would be to see a floating cork suddenly rise and float a half inch higher in a pan of water. To date, the only hypothesis to explain the upward motion of large regions like the Colorado plateau is that of convection currents. Slowly moving convection currents in the solid rock, some 40 to 50 kilometers below the surface of the earth, are presumed to have swept a great volume of light rock from some unidentified place and to have deposited it underneath the Colorado plateau. A total volume of approximately 2,500,000 cubic miles of sialic rock is necessary to account for the uplift of the Colorado plateau. While it is not hard to visualize rocks as having no great strength at the high pressures and temperatures existing at depths of 40 to 50 kilometers, it is quite another matter to visualize currents in solid rock of sufficient magnitude to bring in and deposit this quantity of light material in a relatively uniform layer underneath the entire Colorado plateau region.

The Tibetan plateaus present a similar problem, but on a vastly larger scale. There, an area of 750,000 square miles has been uplifted from approximately sea level to a mean elevation of roughly three miles, and the Himalayan mountain chain bordering this region has floated upward some five miles, and rather late in geologic time, probably within the last 20,000,000 years. The quantity of light rock required to be swept underneath these plateaus by convection currents to produce the effects noted would be an order of magnitude greater than that needed to uplift the Colorado plateau— approximately 25,000,000 cubic miles. Even more troublesome than the method of transporting all this light rock at shallow depths below the surface of the earth is the problem of its source. The region from which the light rock was moved should have experienced spectacular subsidence, but no giant neighboring depressions are known. A lesser but large

CONTINENTS, MOUNTAINS, AND OCEANS

problem is how such enormous quantities of light rock could be dispersed so uniformly over so large an area.

This evidence of uplift and downsinking of various crustal blocks, with the blocks always remaining in approximate isostatic balance, does not seem to harmonize with the view of a floating sialic continent on a denser substratum where one might expect to find little variation in elevation with time.

The second problem, that of the subsidence of troughs, is of equal difficulty. The rivers of the world carry enormous quantities of sediments seaward. Most of this sedimentary burden is deposited within a few tens or hundreds of kilometers of the shore line and little is transported to the deep ocean basins. Thus, elongate prisms of sediments are built up parallel to the shores of certain regions where great quantities of sediments are transported to the sea. The crust, in response to this added load of sediments, begins to buckle downward. Troughs filled with sediments appear, paralleling the coast line. The chicken and the egg argument enters here, for it is not entirely clear whether deposition of sediments generates the troughs or whether the troughs are formed first and are later filled with sediments. However this may be, one such trough now in the making is along the coast of the Gulf of Mexico on both sides of the mouth of the Mississippi River. Surprisingly enough, this trough deepens at about the rate new sediments are added to it. Thus, the sediments are always deposited in relatively shallow water.

Fundamental laws of physics are violated and on a large scale if this downwarping is produced directly by continued loading of sediments. These deep troughs filled with sediments may contain 50,000 to 100,000 feet of sediments and may be 1,000 or more miles long and 100 miles in width. The mean density of the sediments, even compacted under a load of 10,000 feet of other sediments, is approximately 2.4 to 2.5. The rocks displaced in the downwarping trough are known to be denser, with a mean density of 2.8 to 2.9. By what mechanism do light sediments displace denser, crystalline rock? These troughs of sediments, like the plateaus considered earlier, always appear to be in isostatic balance. If the conventional is to be sustained, dense rock must automatically be removed from below the bottom of these sedimentary troughs at approximately the same rate at which they receive sediments from the rivers that feed them, so that the troughs balance and float with their upper layers of sediments under a few tens or hundreds of feet of water most of the time.

The problem of the mechanics of the formation of deep troughs of low density sediments is heightened when their full history is considered. Many are known in the geologic record. In most, sediments accumulate for perhaps a hundred million years and reach a total thickness of as much as 100,000 feet. These thick, highly elongate lenses of sediments may then be slowly folded and uplifted to form mountain ranges which may initially stand as much as 20,000 feet high. Surprisingly, the geologic record shows that a large fraction of the mountain ranges of the world has been formed from rocks of these thick, geosynclinal troughs. Extensive volcanic activity may accompany and continue beyond the time of the formation of the mountain ranges. The mystery then, of the down-sinking of the sedimentary troughs, in which low density sediments apparently displace higher density rocks, is heightened when we note that these narrow elongate zones in the earth's crust, the most downwarped, with the greatest accumulation of rock debris, shed by the higher portions of the continents, become in turn the mountain ranges and the highest portions of the continents.

The third of the major problems connected with the postulated sialic continental area and simatic oceanic region is that pointed out by recent measurements of flow of heat through the crust of the earth.

Considerable numbers of measurements have been recently

made of temperature gradients and rock conductivities within the outer part of the earth's crust. Careful temperature profiles have been made within many of the accessible deep mines and in numerous wells and tunnels. From these data a fairly reliable picture has developed of heat flow within the earth's outer crust, although measurements are not nearly so detailed or so numerous as could be desired. The rate of escape of heat through most continental areas appears to be approximately 1.2 microcalories per centimeter per second. It has been known for many years that most of the heat escaping from the earth is radiogenic heat, generated in the earth by decay of radioactive isotopes of uranium, thorium, and potassium. Little or none of the escaping heat is primary, inherited from an initially hot earth. In fact, there is no compelling evidence that the earth was molten in its youth or even was formed from hot material. We know that the rocks near the surface today appear to be in fairly reasonable thermal balance. The rate of heat escaping from them to the surface of the earth is very close to the rate at which heat is generated in them by radioactive decay of certain elements.

Over the last twenty years, extensive data have been accumulated concerning the distribution of the radioactive elements. Uranium, thorium, and potassium are 10- to 100-fold as abundant in the light silica-rich rocks as they are in denser simatic material, rich in magnesium and iron, and low in silica. Consequently, we might expect that radiogenic heat in the thick sialic continents would be vastly greater than the heat generated in the presumably radioactive-poor simatic material underlying the ocean floors. Further, we would expect heat flow to be greatest in the thickest parts of the continents; that is, in mountainous regions buoyed up by thick roots of sial rich in radioactive elements. Many studies of heat flow through the continents have been made over the last two decades by examining the distribution of temperatures and rock conductivities down deep wells and along tunnels. Surprisingly, these studies show almost no correlation between mean elevation of land mass and heat flow through the earth's crust. This was most unexpected, because all the broader regions of higher elevation are presumably underlain by thick zones of light rock which, from all determinations, should be richer in radioactive elements.

Nonetheless, it was confidently expected that heat flow through the floor of the ocean would be a fraction of that observed in the continental land masses. The first measurements of heat flow through the floors of the ocean were reported in 1952 by Sir Edward Bullard [1]. These determinations were ingeniously made by inserting probes containing thermisters into the muds on the ocean floors. Startlingly, the heat flow determined by these measurements through the floor of the ocean was almost identical with that measured in continental and mountainous areas. Later results by Revelle and Maxwell ([2] and unpublished), although indicating wide ranges of heat flow from place to place in the oceans, have only affirmed the earlier observation that heat flow through the ocean floor is essentially the same as that on the continents.

There seem to be only two possible explanations for this most unexpected discovery: either the concentration of radioactive elements in the rocks below the floor of the ocean is the same as that in rocks which make up the continents or else heat is transferred by some special mechanism from deeper in the earth to near-surface sites underneath the oceans. If the concentration of radioactive materials in the few tens of miles below the floor of the ocean is the same as that in a few tens of miles below the continents, then our previous view that the ocean floors are underlined by radioactive-poor sima and the continents are underlain by radioactive sial certainly cannot be right. The alternative explanation, equally difficult, is that high-temperature rocks from deeper in the earth are convectively carried up to near-surface environments below the oceans. Thus, heat escape through the floor of radioactive-poor oceans fortuitously approximates heat escape through the radioactive-rich continents.

The fourth problem, that of the long lifetime of continents and mountain ranges, is perhaps the most difficult of all. The rivers of the world strip tremendous quantities of rock debris off the continents each year and deposit it in the oceans. The Mississippi, for example, contains about one-half weight per cent of solids as it flows into the Gulf of Mexico. Each year it brings to the Gulf of Mexico approximately 750 million tons of dissolved and solid material. The great rivers are steadily wearing down their basins. Calculations show that the Missouri River lowers its drainage basin about one foot in each 8,000 years, and that the rate of erosion for the entire United States approximates one foot in 10,000 years [3]. At this rate all the land masses of the world would be eroded to sea level in something of the order of 10–25 million years. This is particularly surprising in view of the fossil record. Land animals and plants have been known on the surface of the earth for well over 300 million years, and the sedimentary record indicates high land masses extending back at least 2 billion years. Much geological evidence indicates that the ancient continents were in approximately the same place as the present continents and that continents have existed more or less as they are today and for a period of at least 2 billion years. How do we reconcile an erosional lifetime for continents of something like 25 million years with a known lifetime of something of the order of 2 billion? Why has not all the continental sial been uniformly distributed through the ocean basins?

The mountain ranges bordering the continents and interior to the continents present an even more difficult problem. The rates of erosion along the slopes of steep mountains are many times those of lower-lying continental land masses.

The lifetime of mountains, therefore, must be far less than the 25 million years estimated for continents. In contrast to this reasoning, however, is the geologic record which strongly suggests that the Appalachian Mountain Range has existed more or less where it is today and, as far as we know, with reasonably similar relief for the last 200 million years, shedding sediments both to interior valleys and coastward. Thus, we see orders of magnitude discrepancy between lifetimes of mountain ranges and continents, estimated on the basis of known rates of erosion, and the lifetimes of the mountains and continents as indicated by the geologic record. Even though we assume that mountain ranges and continents are somewhat analogous to icebergs that float up as their exposed portions are melted away, the presumed depth of roots of the mountain ranges and thickness of the light continental rocks permit extension of the estimated lifetime of continents by no more than tenfold that based on present erosion rates and mean elevations.

Thus again, the notion that the rocks which make up the continents are grossly different in composition from those underlying the ocean basin does not seem to hold up, for we would expect that the rain waters washing over the continents would have long ago dispersed the continental rocks into the oceans.

These four major observations then—persistence of continents and mountain ranges in spite of high erosion rates, the relatively uniform values for heat flow in continents and ocean basins, subsidence of marginal troughs in response to loading by low density sediments, and uplift of plateaus once worn to sea level—suggest the inadequacy of the traditional view that continents represent masses of low density silica and alumina-rich rock floating in the denser media of sima, iron, and magnesium-rich rock.

Recent theoretical studies by MacDonald [4] and experimental work by Robertson et al. [5] and by the writer, as well as interpretation by J. F. Lovering (1958) [6], suggest a different structure of continents, a structure which simultaneously explains most of the observed phenomena associated with continents, mountain ranges, and ocean basins and accounts for the four major stumbling blocks in existing theory. This new model of the earth's crust stems from theoretical considerations largely confirmed by recent experimental work in the field of high pressures.

Very many crystalline solids undergo polymorphic inversions to denser phases when subjected to high pressures. The behavior of matter at high pressures has been extensively investigated by Bridgman [7], who has demonstrated literally hundreds of polymorphic inversions among common substances in the pressure range 0–100,000 atmospheres. Graphite and diamond form, for example, a familiar polymorphic pair. At sufficiently high pressures and temperatures graphite may be converted to diamond. A temperature of 1,500 K and a pressure of 100,000 atmospheres is sufficient for the conversion, and, indeed, many thousands of carats of diamonds are now being made annually by General Electric Company by subjecting carbonaceous material to high temperatures and pressures [8].

It has long been noted (see, recently, MacDonald [4]) that basalts and eclogites, rocks with sharply contrasting mineralogy, have essentially identical chemical composition (see Table 1).

TABLE 1

	Eclogite [4]	Plateau Basalt [9]
SiO_2	48.12	48.80
TiO_2	.85	2.19
Al_2O_3	10.42	13.98
CaO	9.99	9.38
MgO	14.22	6.70
FeO	13.92	13.60
Na_2O	1.45	2.59
K_2O	.58	.69

Eclogite, however, contains no feldspar; instead, it is made up of jadeitic pyroxene and garnet. The mean density of eclogite is 3.3 g per cc, that of gabbro is 2.95 g per cc. As eclogite is the denser of the two phase assemblages, it is the rock that must exist at the higher pressures.

The density contrast, about 10 per cent, between gabbro and eclogite is almost the same density contrast believed from seismic evidence to exist at the M discontinuity, although the contrast at the discontinuity has usually been assumed to be a chemical rather than a phase contrast.

Indeed, Fermor [10], Holmes [1], and Goldschmidt [12] suggested that M discontinuity might be a phase contrast and that the rocks below it are eclogite. Their suggestion received little discussion or acceptance but has been recently revised by MacDonald on the basis of calculations of the pressure-temperature conditions controlling the phase change of nepheline plus albite to jade and of albite to jade plus quartz. The calculations of MacDonald [13] were based on new thermochemical values for heat capacity at low temperatures and heats of solution of nepheline, albite, and jade by Kelley and his colleagues [14]. Similar calculations [14,15] have firmly established the slope of the transition in a pressure-temperature plane of the reaction, nepheline plus albite = 2 jade, and that of the reaction, albite = jade plus quartz.

These thermochemical calculations have been confirmed by experimental work of Robertson et al. [5] and by the writer. These two experimental studies, though in disagreement in detail, confirm the calculations based on thermochemical data that, at pressures of 15,000 to 25,000 atmospheres, depending on temperature, the nepheline plus albite undergoes a polymorphic change to jade, and albite undergoes an inversion to jade plus quartz at slightly higher pressures. Further, an experiment made by me on basalt glass showed that, at 500° and pressures below 10,000 bars, basalt glass crystallizes as gabbro. The major mineral component is feldspar. At pressures above 10 kilobars and at a temperature of 500°, the amount of feldspar decreases and, finally, basalt glass crystallizes directly to a rock made up dominantly of jadeitic pyroxene. Identification of phases were by X-ray. Significantly, 500° and 10 kilobars are approximately the temperatures and pressures estimated at the M discontinuity underneath the continents. It thus appears that the M discontinuity may reflect a phase change from gabbro to eclogite rather than a change in chemical composition. This phase change will account for the observed change in seismic velocity from approximately 6.5 to 8.1 kilometers per second and a change in density from 2.9 to approximately 3.23. Thus, the older suggestions of Fermor, Holmes, and Goldschmidt are supported by field measurements, theoretical calculations, and recent experimental work.

If the discontinuity caused by a phase change takes place at a depth of 30 kilometers, a depth equivalent to a pressure of approximately 10 kilobars under the continent, how do we account for the much greater depth to the discontinuity under mountain ranges and the much shallower depth to the discontinuity under the oceans? The answer lies in the fact that the change takes place at a different pressure for a different temperature. As near as can be told from the computations and from the experimental data, the slope of this phase change is approximately the same as the earth's pressure-temperature gradient, as indicated in Figure 2.* Consequently, if it is assumed that the earth's temperature increases a little more rapidly per foot of depth under mountain ranges than under continents generally, the transition will take place at a vastly greater depth (depth C in Fig. 2). If it is assumed that the earth's temperature increases with depth a little more

* The pressure-temperature gradients of Figure 2 are approximately the same as those computed on the assumption that mean heat flow is approximately 1.2×10^{-6} cal/cm/sec and that half the heat is radiogenic heat, generated in the upper 40 kilometers of crust. The remaining half is from below.

slowly under the oceans than under the mountains and continents, the transition is at shallow depths (depth A). Thus, the single transition explains the varying depths to the M discontinuity under the oceans, mountain ranges, and continents.

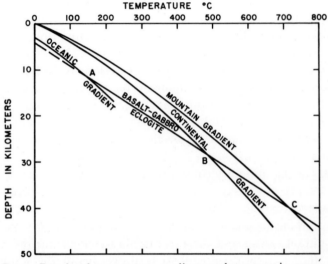

FIG. 2. Postulated temperature gradients under mountain ranges, continental areas, and oceanic regions.

We assume that there are variations in temperature from continents to ocean basins to mountain ranges, and, consequently, we would expect variations in heat flow. However, the necessary variations in heat flow to account for these different depths of intersection are exceedingly small, well within the range of observations and are certainly not the threefold variations in heat flow that we would expect if the continents and mountain ranges were thick zones of radioactive-rich sial and the ocean was underlain by radioactive-poor sima.

It is interesting to note in Figure 2 that, within the assumptions used in drawing this graph, the earth's pressure-temperature gradient is almost the same under the oceans as is the slope of the phase change. The intersection here is assumed to be at low pressures and temperatures (point A in Fig. 2). Because the temperature is very low, reaction rate of the phase change might be expected to be very slow, and the response of the discontinuity position under the oceans might be extremely sluggish to small changes in temperature and pressure. Thus, we may not always have thermodynamic equilibrium under the oceans.

Early in this discussion it was noted that the relief of the earth's crust is a direct function of the thickness of the zone of light rock. If the thickness of the zone of light rock reflects the depth of the M discontinuity, which it almost certainly does, the relief of the earth's crust can be interpreted as mirroring the various temperature gradients in the upper part of the mantle.

The four major problems of the surface of the earth, discussed earlier, seem satisfactorily explained by phase transition. A chemical contrast at the discontinuity is unnecessary. The rocks on both sides of the M discontinuity may thus be of the same composition, and the depth to the discontinuity may be a function of very slight temperature variations from place to place in the earth's crust.

The uplift of continents, once at sea level, to high plateaus would be a consequence of warming the rocks near the M discontinuity a few tens of degrees. When this happened, the phase change would migrate downward to much greater depths. The dense rock below the discontinuity would become light rock and the volume increment would float the continents to higher levels. Thus, convection currents are no longer needed to transport millions of cubic miles of light material underneath the continents in order to float them higher into the air.

Similarly, the long lives of mountain ranges are explained. As the tops of mountains are eroded away, pressure at the discontinuity deep below the mountains decreases. Dense rock at the discontinuity would be converted to light rock, so light roots underneath the mountains would be recreated to keep them floated to high elevations.

The downsinking of sediments in troughs is also explained by the phase transition. If sediments from a mountain range were rapidly removed and deposited in troughs, the first effect of loading would be to increase the pressure at the base of the trough with very little change in the temperature. Consequently, the discontinuity would migrate toward the surface. The trough would sink, not only because of the added load of rock at the surface, but because light rock would be converted into dense rock at the discontinuity below the trough, with a consequent decrement in volume of material below. Thus, the short-term effect of rapid sedimentation is one of sinking. A most interesting long-term effect appears. The added new sediments filling the trough are of low thermoconductivity and possibly richer in radioactive material than the surrounding rock. Consequently, given sufficient time, the temperature would slowly rise at the bottom of the trough and, although the discontinuity would first migrate surfaceward under response to loading, it would ultimately migrate downward under response to the rise in temperature owing to the blanket of poorly conducting sediments rich in radioactive elements deposited in the trough. Thus, troughs might sink for considerable time and then be uplifted to form mountain ranges as the roots of the trough deepen with warming of the base.

This implies that mountains are generated largely because of vertical motion and not lateral thrust. A good deal of the faulting and folding of rocks in mountain ranges is assumed to be the result of load. By this thesis, the major folds and faults associated with mountain chains are gravitational in origin, although concomitant lateral thrust of other origin is not excluded.

The problem of the relatively uniform heat flow to the surface of the earth is readily explained by the phase-transition concept. The earlier crustal models assumed that continents were made up of silica-rich and radioactive element-rich rocks. Thus, continents should, but do not, show heat flows several times that of oceanic areas. If the bulk composition of continental rocks were not vastly different from the bulk composition of oceanic rocks, we would expect relatively uniform heat flow from place to place in the earth's crust. This is indeed what we do find. The precision of measuring heat flow, however, is not sufficiently great to exclude the possibility that minor variations in temperature do exist from place to place in the earth's crust. In fact, it is necessary to appeal to these minor variations to account for the existence of ocean basins, mountain ranges, and continents on the basis of a phase change as discussed here.

If we assume the M discontinuity to be a phase change, many questions are answered, but other questions are also raised. The phase change cannot be a simple solid-solid phase change inasmuch as the major minerals involved are of variable composition. Consequently, the change must take place over a considerable depth interval and should not be a sharp change taking place at a fixed depth. The data of seismology bear on this problem. They permit the interpretation that the discontinuity may take place, instead of at a given depth, over an interval of as much as 10 kilometers under the continents (Frank Press, oral communication). This is within the requirements of the change. However, more difficult problems emerge when oceanic areas are considered. The discontinuity under the oceans is very shallow and apparently takes place over a very narrow depth interval. In fact, the pressure interval seems much too narrow to represent the gabbro-eclogite change. However, further experimental work needs to be done to measure precisely the required pressure interval, and more refined seismic work will be necessary before we know exactly the distribution of seismic velocities below both the oceans and the continents.

ACKNOWLEDGMENTS

Gordon J. F. MacDonald first brought to the writer's attention the suggestion that the Mohorovičić discontinuity was a phase change. The experimental confirmation of the reality of the phase change in the laboratory would not have been undertaken without his stimulation. D. T. Griggs has contributed much to the author's understanding of the problems involved. G. D. Robinson, V. E. McKelvey, and D. M. Hopkins have critically reviewed this manuscript and many thanks are due them.

REFERENCES

1. BULLARD, E. C. Heat flow through the floor of the eastern North Pacific Ocean. *Nature*, **170,** 202 (1952).
2. REVELLE, R., and A. E. MAXWELL. Heat flow through the floor of the eastern North Pacific Ocean. *Nature*, **170,** 199 (1952).
3. GILLULY, J., A. C. WATERS, and A. O. WOODFORD. *Principles of Geology*. San Francisco, W. H. Freeman (1952).
4. MACDONALD, G. J. F. Chondrites and the chemical composition of the earth. In: *Research in Geochemistry*, P. H. Ableson, Ed., New York, Wiley (1959).
5. ROBERTSON, E. C., F. BIRCH, and G. J. F. MACDONALD. Experimental determination of jadeite stability relations to 25,000 bars. *Am. J. Sci.*, **255,** 115 (1957).
6. LOVERING, J. F. The nature of the Mohorovičić discontinuity. *Trans. Am. Geophys. Union V.* **39,** 947 (1958).
7. BRIDGMAN, P. W. *The Physics of High Pressure*. London, G. Bell and Sons (1952).
8. BUNDY, F. P., H. T. HALL, H. M. STRONG, and R. H. WENTORF. Man-made diamonds. *Nature*, **176,** 51 (1955).
9. DALY, R. A. *Igneous Rocks and the Depth of the Earth*. New York, McGraw-Hill (1933).
10. FERMOR, L. L. The relationship of isostasy, earthquakes, and volcanicity to the earth's infra-plutonic shell. *Geol. Mag.*, **51,** 65 (1914).

11. HOLMES, A. Some problems of physical geology in the earth's thermal history. *Geol. Mag.*, **64**, 263 (1927).
12. GOLDSCHMIDT, V. M. Ueber die Massenverteilung im Erdinneren, vergleichen mit der Struktur gewisser Meteoriten. *Naturwissenschaften*, **10**, 918 (1922).
13. MACDONALD, G. J. F. *A Critical Review of Geologically Important Thermochemical Data*. Doctoral dissertation, Harvard University (1954).
14. KELLEY, K. K., S. S. TODD, R. L. ORR, E. G. KING, and K. R. BONNICKSON. *Thermodynamic Properties of Sodium-Aluminum and Potassium-Aluminum Silicates*. U.S. Bureau of Mines Report of Investigations 4955 (1953).
15. ADAMS, L. H. A note on the stability of jadeite. *Am. J. Sci.*, **251**, 299 (1953).

WHITE DWARFS AND STELLAR EVOLUTION

By Willem J. Luyten
University of Minnesota

About fifty years ago astronomers were beginning to feel comfortably sure that they understood what stars are, how they are born, how they change and die. We had just found our key diagram in astrophysics, the famous Hertzsprung-Russell diagram which gave the relationship between luminosity and surface temperature of a star and according to which all stars appeared to be arranged in a practically continuous sequence. At the end of it lay the very cool, very red stars of low luminosity—the dying embers of the stellar population. But just then came the discovery of two stars of very low luminosity but *high* surface temperature. At first the reaction was: "there ain't no such animal," somebody must have goofed. When we realized they were real, that something new has been added to the known species of stellar populations we also quickly realized that this new species was destined to play an important part in our theories of stellar and atomic structure and of stellar evolution.

So important has this discovery become that we should perhaps briefly summarize the observational history behind it. Ever since the middle of the nineteenth century it had been known that Sirius, the brightest star in the sky, moved in an oscillating manner, the cycle repeating itself in about 50 years. Now, no self-respecting star is supposed to move in anything except a straight line during so short a time: oscillations and curves mean acceleration, that in turn means a force, and to the astronomer this means gravitation. In no time at all Auwers had proved that Sirius had a faint companion and that the two stars were revolving around each other in 50 years. In a sense, therefore, this faint companion was discovered—with the eye of gravitation—before it was *seen*. Then, in 1862, while testing the lens which now is in the telescope of Northwestern University, Alvan Clark, Jr. looked at Sirius and saw the companion, a very faint star, almost exactly 10,000 times fainter than Sirius itself, and very close to it. Difficult to observe, certainly, but by 1910 we had enough reliable data to calculate a good orbit. The distance to Sirius is accurately known—8.7 light-years—and from all this we could piece together the information that the faint companion has a mass equal to 0.96 of that of the sun, but is 400 times less luminous. Nothing unusual thus far, but then came the bombshell: in 1915 Adams, at Mt. Wilson, observed the spectrum and found the star to be hotter than the sun, with a surface temperature around 8,500°K. We are so convinced of the correctness of our laws of radiation that we immediately used this temperature to calculate that each square inch of that faint companion must give three times as much light as the sun. But with a total luminosity of $1/400$ this would imply a surface 1,200 times smaller than that of the sun, a diameter of $1/35$ and a volume of $1/42,000$ of that of the sun. Finally, with a mass practically equal to that of the sun the density comes out to more than 50,000 times the density of water, or almost a ton per cubic inch! Nothing like this had ever been observed before, yet the observational data were checked and rechecked—they were essentially correct. Recent work may have changed these values somewhat, here and there, but the final result is a density even higher—almost unbelievable.

WHITE DWARFS AND STELLAR EVOLUTION

The second star of this kind, one of the components of a faint double star in the constellation Eridanus, was found by Harvard astronomers at about the same time, and a third, similar object was found by Wolf of Heidelberg, and Van Maanen of Mt. Wilson very shortly afterwards.

The explanation, by Eddington, was not long in forthcoming. It lies in our understanding of the nature of the atom. In the interior of these stars all atoms may be completely ionized, stripped of virtually all their electrons and reduced to what is now called degenerate matter. Sir Oliver Lodge has given perhaps the most dramatic description of it. Under ordinary conditions matter may be compared with flies buzzing in a cathedral; shell after shell of tiny electrons revolving around the tiny nucleus of the atom, in orbits tens of thousands of times larger in diameter than the particles themselves. The orbits represent the cathedral walls, the electrons and the nucleus the flies. But in these stars we are not dealing with ordinary matter. Inside these stars, at the fantastically high temperatures and pressures that exist there, the electrons fly off the handle. The cathedral walls collapse and all there is left is the flies. One can pack a lot more flies than cathedrals into a city square.

Because these stars are so small in size, comparable with planets rather than ordinary stars, and are, in general, of such high surface temperatures that they appear blue or white in color, they have been named "white dwarfs." Not only their densities are unbelievably high but also their surface gravities. If a man could survive on the suface of the star, then someone weighing 150 pounds on earth would there weigh some 25,000 *tons*. One consequence of that Eddington immediately predicted. If relativity is correct, then such a high value of gravity must slow down the vibrations of light rays leaving the surface of such a star and the lines in the spectrum should be shifted toward the red. Adams observed the spectrum of the Sirius companion very carefully; while the observations are extremely difficult and we do not have even yet a really accurate answer, he did obtain a "red-shift" of the right order of magnitude. Perhaps it is little realized today that this killed two birds with one stone in that it not only proved Eddington's ideas about the structure of white dwarfs to be correct, but it also provided one of the only three experimental verifications of relativity then possible and, as such, it aided greatly in the general acceptance of Einstein's Theory of Relativity.

So now we had three white dwarfs. What next? Obviously the next thing to do was to find more of them, to see how many varieties there are, how common or uncommon they are in space, how they got that way, and what eventually becomes of them. As to what they *are*, the theoreticians got to work, and after Eddington came Fowler and Chandrasekhar who soon made a thorough investigation of the properties of matter under these extreme conditions. The long and short of it is that quantum principle demands serious deviations from the classical gas laws under these conditions. It is then that we begin to speak of "degenerate matter" and the conclusion of Chandrasekhar's work was that such highly degenerate matter provided a good first approximation for the structure of white dwarfs.

Personally I am not a theoretician and my own interests and work have been concerned mainly with the observational aspect of white dwarfs, so I will concentrate on that. In 1921 I began to observe the spectra of stars of low luminosity; out of 100 stars observed only one proved to be a white dwarf and even that one was not generally so recognized until many years later. It became evident that most white dwarfs were beyond the reach of the then existing spectroscopic equipment, and furthermore that, if more of them were to be found, we must first discover more stars of low luminosity.

For the next 20 years I concentrated mainly on that, using plates from the Harvard Observatory. The first thing we had to do was to repeat the old plates—this was begun in 1928 and completed by the Harvard Observatory in 1935. Then the old and new plates were compared in a "blink microscope" which enables one to pick out the stars that have moved appreciably in the interval. In this way the first "screening" selected some 100,000 stars of appreciable motion from among the 30-odd million stars shown on the plates. Now, a star that moves rapidly across the sky—in angle—must be *near*, for otherwise it must possess an enormous linear velocity and such stars simply do not occur. If a star is near, and also very faint in appearance it must be a star of low luminosity. Thus, by selecting from among the 100,000 stars with appreciable motion the 5,000 very faint stars with the largest motions we have effectively selected the 5,000 intrinsically faintest objects among the original 30,000,000. In other words we have the real dwarfs, and these constitute the richest potential source of white dwarfs. The only thing left to do is to see what color these stars have. To do that we simply take another pair of plates, one in blue, one in red light. As expected, the vast majority of these 5,000 stars turned out to be ordinary red dwarfs, of no further immediate interest, but we did end up, in this way, with a few hundred faint white stars with large motions, almost certainly white dwarfs. Simple, is it not? Yes, but an awful lot of hard work.

The first part, the screening out of the 100,000 moving objects, and then the further screening out of the 5,000 best prospects from among the original 30,000,000 I did at Minnesota and it took roughly ten years. The second part was actually much simpler—but more difficult to execute. We have no observational equipment at Minnesota with which to take these color plates and, for a number of years, I was unable to interest any other observatory in the program. It was Carpenter at Arizona and Gaviola at Cordoba who gave me a chance to begin the color observations. Once the first new white dwarfs came out of the mill, one could again observe the truth of the old adage that nothing succeeds like success. By 1947 the observational program was rolling along beautifully and I quickly graduated into the most successful parasite among contemporary astronomers, using telescopes, or getting plates taken for me at a dozen different observatories.

In the method sketched here the motions are determined first, the colors afterward. In the early forties, Zwicky and Humason at Mt. Wilson started at the other end. Their ingenious idea was to look for faint blue stars in regions where distant—and therefore luminous—stars are not expected to occur, such as in heavily obscured regions (the Hyades), where any blue star found must lie in front of the cloud of obscuration, or near the Galactic Poles where we look through the thin part of our stellar system and do not expect to find stars at great distances. In 1947 Zwicky and Humason published their first list of 48 blue stars found in this manner. Those in the Hyades proved, indeed, to be mainly white dwarfs, but among the 33 blue stars found near the North Galactic Pole only one or two appear to be white dwarfs. The logic behind this research was flawless but, as Kettering once said, "logic is an organized way of going wrong with confidence." In science, however, such going wrong often means making a new discovery and the Zwicky-Humason stars, as these objects are now generally called, proved to be the forerunners of another new and important type of stellar object.

Following Zwicky and Humason, further searches for this type of blue star have been made at Minnesota and at Tonantzintla by Haro, Iriarte, and Chavira. Altogether we now have some 5,000 objects of this kind, comparatively few of which will prove to be white dwarfs—the great majority being "Z-H" objects—but of them more later.

Gathering the results from all different approaches we now have around 400 white dwarfs, certain, probable, and possible, more than 330 of which were found at Minnesota. To make certain of all of them we need more observations; here we run up against the paradox that while white dwarfs are fairly easy to discover they are extremely difficult to observe. Generally, they are so faint that only the very largest telescopes can get their spectra with sufficient detail; almost no observatories are equipped (or willing!) to determine their distances and it is only within the past few years that improvement in photoelectric techniques has made it possible to observe their colors with sufficient accuracy.

Even so, we have gradually—almost painfully—acquired enough data to arrive at least at some preliminary conclusions. Distances are known for about 25 white dwarfs; for another 20 belonging to clusters or binaries having a second "normal" component we can make a fairly good guess at the distance and when the distance is known we can derive the intrinsic luminosity. Harris at McDonald has determined accurate colors for about 60 and Greenstein at Palomar now has detailed spectra for about 80 of them.

Briefly, what has emerged from this is the picture of white dwarfs as forming a continuous sequence, running from the bluest (and hottest) with luminosities of about $\frac{1}{40}$ of that of the sun to the yellowest (and coolest) with a luminosity of about $\frac{1}{25,000}$ of that of the sun. Not that the sequence stops at either end: at the top it may tie in with the Zwicky stars while at the lower end it runs into stars of a type which are so faint as to be almost observation-proof. In size, the now known white dwarfs seem to range from a diameter of five times greater to three times smaller than that of the earth, the smallest known specimen appearing to be one which I found some years ago, and having a diameter about equal to that of the planet Mercury. When it comes to *mass* the situation is much less satisfactory. Only three masses are known and two of these belong to objects which may be somewhat peculiar and for which the other data such as spectrum, temperature, and color are difficult, if not impossible to obtain. Thus, really, the mass of only one bona fide white dwarf is known but this has not prevented the theoreticians from erecting an imposing structure of formulae, theories, and predictions on it. Thus, in a recent theoretical treatise, the author laments the fact that only two masses are known, and then proceeds to state categorically that the range in mass *is* small and produces two separate diagrams in which the masses of seven more white dwarfs are used, and described as *known* when they are, in fact, calculated from the theory the author is trying to prove. Evidently this author, as so many other theoreticians, belongs to that group of which it could be said, as Justice Holmes said of the Supreme Court: they are "often in error, but never in doubt."

There are now known some 40 wide double stars with one white-dwarf component, and two double white dwarfs. If accurate observations of the relative positions of the components of these binaries are made, some estimates of the orbital motion may be made in another 25 years or so, and from this, in turn, at least some statistical information on the masses may be obtained.

In spite of the uncertainty in the masses we feel sure that the densities of white dwarfs must be very high indeed, probably averaging well over 100,000 times that of water. This is partly borne out by the spectra which usually show very broad and diffuse lines—the broadening generally believed to be due to electrostatic forces between the closely packed nuclei and electrons. Rotation may play a role, although until we have an adequate picture of *how* stars become white dwarfs it is difficult to make quantitative estimates.

With their spectra the strange paradox is that the majority of these stars show only, or mainly, the lines of hydrogen, yet it is now generally believed that their over-all hydrogen content must be less than one part in a hundred thousand—most normal stars are around 50 per cent hydrogen. These

WHITE DWARFS AND STELLAR EVOLUTION

stars are considered to be mainly composed of helium and heavier elements and it is visualized that the "degenerate core" takes up practically the whole body of the star, that this is surrounded by a "mantle" less than 100 miles thick, with finally an atmosphere—in which the hydrogen lines originate—only a matter of a few hundred feet thick. There are even speculations to the effect that the small amount of hydrogen now present is really foreign to the star, and may have been acquired by accretion from gas clouds in space.

If they contain virtually no hydrogen, it follows that these stars cannot produce energy—i.e. "shine"—by converting hydrogen into helium and, as Mestel has suggested, white dwarfs may be the only species of stars that "live" by the Helmholtz contraction theory, converting gravitational into luminous energy.

One further interesting speculation may be added. We could obtain an enormous amount of information about white dwarfs if only we could find a pair of them revolving so closely around each other as to produce eclipses but, unfortunately, the chance of finding such a pair, even if they exist, is remote. Imagine two stars each the same size as the earth and with half the mass of the sun, revolving around each other at a distance of three earth diameters; the period of revolution would be around two minutes, each eclipse would last only 10–15 seconds, and their orbital speed would be around 1,250 miles per second. Chances would be a good deal better for a binary containing one normal component with, say, a diameter equal to one-third that of the sun, and where the mutual separation would be equal to the sun's diameter. The period then would be something like eight hours and totality in the eclipse would last half an hour. We have one possible suspect for this, a star in the Hyades, but the evidence thus far is only circumstantial and not enough to convince.

Before we begin consideration of where these stars fit in to the general scheme of things we must first mention Baade's discovery, 15 years ago, that there appear to be two different kinds of stars in our universe. He designated these as Population I and Population II, the first group conforming to a Hertzsprung-Russell diagram similar to that of the stars in the vicinity of the sun, the second group conforming to that of stars in globular clusters. The question of age also enters into this. Population II stars seem to be the old, primeval stars and some globular clusters are estimated to be seven billion years old. Very hot, very blue, very luminous stars are "extreme Population I" stars and these stars are so prodigal with their hydrogen fuel that their life expectancy is very short, and so we do not find this kind of object in an old globular cluster.

We now believe that the sun has existed for some 5 billion years or so, and in pretty much the same form as it is now, and that its hydrogen fuel reserves are such that, if no changes take place in its fuel consumption (but this is a pretty big IF), it is good for at least another 20 billions years. One of these superluminous aristocrats of space, 100,000 times as luminous as the sun but perhaps only thirty times as massive, would squander its entire fuel supply in a few million years. So we conclude that these blue stars must be very young stars, that they were formed only yesterday, astronomically speaking, out of the vast clouds of interstellar gas, mostly hydrogen, that pervade large parts of space. Lo and behold! We now find that these same bright blue stars are practically always associated with such interstellar clouds and that in other galaxies these same blue stars are found along the spiral arms—where the gas clouds are—whereas the bright red stars of Population II are all over the disc of a galaxy and probably make up part of the spherical corona surrounding a galaxy.

The distribution of globular clusters as well as their motions indicate that they are very old structures but the so-called galactic clusters—the Pleiades, Hyades, Praesepe, and so forth—are probably much younger. The Pleiades cluster still has blue "supergiants" and seems to be the youngest of these, the other two no longer have blue stars and seem to be older.

This, of course, gives only the briefest and most oversimplified outline, and it is already apparent that a number of other factors such as total mass, chemical composition (i.e. the fraction of heavy metals present in a star), axial rotation, and so on, enter into the picture of the speed with which stellar evolution proceeds. On the whole, however, it looks as if an over-all, consistent, unified picture of stellar evolution is at last beginning to take shape.

To come back to our original subject: Where do the white dwarfs fit in? Ever since these were discovered there have been recurrent hints and suggestions that they are formed after a star blows up and becomes a nova or a supernova. This agrees well with the idea that no white dwarf should have a mass greater than 1.4 times that of the sun. Hence, if a very massive star is to become a white dwarf, it must first get rid of its excess mass and it might do so catastrophically, by exploding. However, it is now believed that an ordinary nova which, at the moment of its greatest glory, reaches a luminosity around 10,000 times that of the sun, does not lose more than a small fraction of its mass in the explosion. Therefore, if it is to become a white dwarf, a star might have to explode many times. Does it? Maybe. We do know "recurrent" novae, and at least one of them, the star WZ Sagittae is now a white dwarf. It blew up in 1913 and again in 1946 becoming, probably, only ten times as luminous as the sun, then lapsing back into its normal, feeble light, $\frac{1}{200}$ of that of the sun. Too feeble, really, to be a regular nova, not feeble enough to be called a regular white dwarf—but maybe it is working hard at getting there. We know a number of "planetary nebulae," thin gas clouds with an extremely hot star at their center, which is usually fainter than the sun, but brighter than a typical white dwarf. At least one of them, the famous Crab nebula in Taurus is definitely a remnant of the supernova of 1054 A.D.

As you can see, we are not sure yet; but the evidence is slowly accumulating and we are beginning to feel sure that, in whichever way they get there, white dwarfs mark the end of the line. They represent the last stage of a star's decline into obscurity and oblivion. [Since I have spent the past forty years trying to discover and observe more white dwarfs, my friends (or are they?) call me a stellar mortician.]

If this picture of white dwarfs is true then we should find few of them among Population I stars and most of them in Population II. This seems to fit in with the observations, although the companion to Sirius, the very first white dwarf discovered, is somewhat puzzling since Sirius is considered to be a rather *young* Population I star and the companion appears to be one of the *older* white dwarfs.

We must also be rather careful not to place too much credence in our many current, but often shaky, theories. In a recent analysis of the situation in clusters Sandage calculated, from an estimate of their ages, the number of white dwarfs that should be present in several galactic clusters and concluded that "the numbers are in good agreement with the observations when account is taken of the incomplete observational search." Now the facts are as shown:

Cluster	Calculated	Observed
h-χ Persei	0	0
Pleiades	2	0
Coma	9	0
Hyades	23	7
Praesepe	20	2

It would seem to me that, unless we can make some quantitative estimate as to *how* incomplete our observational searches are, such a comparison proves nothing as yet.

The same writer mentions that "various authors estimate

that about 10 per cent of the stars in the solar neighborhood are white dwarfs." These high estimates all come from the theoreticians. An estimate I made recently yields a value of only 2 per cent and this estimates is based on a discussion of the motions, distances, and luminosities of 10,000 nearby stars in the southern hemisphere, by far the largest and most homogeneous material now available.

It all depends on how we define a white dwarf. If we include degenerate stars beyond our present limits, i.e. brighter than $\frac{1}{40}$ and fainter than $\frac{1}{25,000}$ of the sun's luminosity, we might be able to raise this value of 2 per cent. However, the former stars are fairly blue, not too difficult to find, and, as it now appears, statistically unimportant in numbers, whereas the latter stars are almost discovery-proof at present and the only indication of their alleged existence in large numbers comes again from pure theory. Here we might well pause and reflect on William Whewell's profound comment: "The cultivation of ideas is to be conducted as having for its object the connexion of facts, never to be pursued as a mere exercise of the subtlety of the mind, striving to build up a world of its own, and neglecting that which exists about us. For although man may in this way please himself, and admire the creations of his own brain he can never, by this course, hit upon the real scheme of Nature."

Summarizing our present knowledge, inferences, and speculations on white dwarfs and relating all of it to our fundamental diagram in astrophysics—the H-R diagram—we reach the following conclusions:

The white dwarfs now known appear to define a locus in the H-R diagram, running from blue stars $\frac{1}{40}$ of the sun's luminosity to yellowish stars $\frac{1}{25,000}$ of the sun, although the vertical dispersion in luminosity appears to be real and to indicate that there may be a real disperion in the fundamental properties of the stars occupying this locus.

We further suspect that this locus continues first to the upper left, or to bluer and more luminous stars, that it then turns straight up, at a luminosity roughly equal to that of the sun, thereafter to turn to the right, and up; i.e. toward stars more luminous but less blue, ultimately to link up, perhaps, with a group of stars which are often referred to as the "horizontal branch" region. This connecting link between the white dwarfs and the horizontal branch is where we now believe the Zwicky-Humason stars to fit in, so they may, after all, be related to white dwarfs. We do not say, however, that this curved extension of the main locus—first suspected by Vorontsov-Velyaminov—is uniformly or even monotonically populated with stars. If it represents the evolutionary track or tracks of stars, different portions may be traversed at different speeds and hence, where the stars linger, their density may be higher—as that of cars in the tunnels on a turnpike—but where they evolve quickly there may even be gaps in the sequence as Haro has pointed out before.

It seems possible now that our sun will some day follow this path; long before all of its hydrogen has been used up gravitational contraction may set in, this will raise the density and temperature of the central regions and the effect may well be a quick if not sudden expansion to a giant star. After this, there may again come a contraction, and perhaps in a future of from five to ten billion years our sun may have passed through the portal of the horizontal branch into the upper reaches of the white dwarf sequence. But all of this is pure theory, and perhaps no more than pure speculation as yet; tomorrow we may well have an entirely different view of it all.

At the other, the faint end of the known sequence, we believe that evolution will carry the stars to lower luminosities and temperatures, hence to redder and redder colors. But the pace of evolution is now so slow that is is difficult if not impossible to predict how many stars we should find at each succeeding step down.

While we are thus uncertain about the "How" and "When," we feel more confident about the "What." The end of the road is a "black dwarf," the ne plus ultra in smallness and degeneracy, a small lump of matter no longer giving light. And after that: Who knows?

TEMPERATURES WITHIN THE EARTH

By John Verhoogen
University of California

The reason why geologists and geophysicists are interested in the temperatures that prevail within the earth is quite fundamental: geologic processes, occurring today very much as they have for the last three or four billion years, invariably involve an expenditure of energy that is probably mostly of radiogenic origin. The manner in which it may become available for geologic processes depends on the intensity and distribution within the earth of radioactive and other heat sources, and it is these two factors, intensity and distribution, that we are trying to evaluate. Obviously, more could be learned about them if we knew exactly what the temperature is now everywhere in the earth.

The dissipative geologic processes just mentioned are of several kinds, the most obvious of which is volcanic activity. In brief, this consists of the outpouring onto the surface of great masses of molten rock (magma), carrying with it large quantities of water vapor, carbon dioxide, sulfur compounds, halogens, and so forth. This material reaches the surface at temperatures of the order of 900–1,200°C, depending on composition; for every gram of lava that cools and solidifies about 400 calories are radiated into the atmosphere and from there into space. The amount of material erupted is of the order of 3×10^{15} g per year at the present, the estimate being uncertain because of our ignorance of the frequency and magnitude of eruptions on the deep-sea floor. Not all magma reaches the surface; erosion reveals in many places "intrusive" rocks of all ages that have obviously risen in the liquid state and slowly congealed at a depth of a few hundred meters or a few kilometers below the surface. In addition, heat is also carried to the surface in volcanic areas by hot springs. All told, some 10^{18}, possibly as much as 10^{19}, calories are dissipated annually in this way.

Erosion is vigorously engaged in leveling all relief, removing material from high places and carrying it to low ones (generally the sea floor); at the present rate, which is possibly typical of all geologic time, it takes somewhere between two and ten million years, depending on climatic conditions, to reduce any mountain range or continent to half its original elevation. In very few multiples of this span of time, the whole surface of the continents would be reduced to a monotonous low-lying plain, were it not for diastrophism, the process that creates mountains about as fast as they are destroyed. The uplift involved requires a considerable expenditure of potential energy, which is again ultimately dissipated as heat in the turbulent motion of the streams that eventually wash every bit of mountain down to the sea. The average elevation of the continents is about 800 meters; to maintain this average elevation against erosion requires an output of 10^{17}–10^{18} cal per year.

A careful study of mountain ranges reveals that they are not merely blocks that have been pushed up; in most instances one sees evidence of intense deformation by which originally flat-lying layers of sedimentary rocks have been tightly folded; great piles of material have slid, or have been thrust upon each other, thereby considerably increasing their initial thickness. A sheet of paper occupies less area on a table when crumpled than when smooth. This gives the impression that the main process involved in mountain making is a shortening, or shrinking, of the earth's crust, and the cause of mountain making has accordingly been sought in the compressive stresses that would develop near the surface of a cooling, and thereby contracting, earth. Geologists are no longer sure that this shortening is any more than apparent; and for horizontal compression as a prime mover they tend more and more to substitute vertical displacements. The matter is of some interest in the present connection, as the contraction theory of mountain making is predicated on the assumption that the earth is cooling. Whether it is or not obviously depends on the heat sources inside.

The mechanical energy dissipated as heat during deformation is probably not very considerable (figures of a few calories per gram of rock have been quoted for the Alps by Goguel) [1], but mountain making is very generally connected with intense metamorphism. By this we mean that the original sedimentary material (shale, sandstone, or whatever) of which most mountains are made has thoroughly recrystallized. Many chemical reactions, most of which are strongly endothermic, have taken place as a result of gradually increasing temperatures which may occasionally reach the melting point (as influenced by chemical composition, presence of water, pressure, and so on). The heat that must be supplied to do this is quite considerable, many times greater than the energy of deformation itself, suggesting that, for all their mechanical appearance, mountain-making processes are essentially thermal. The reverse (exothermic) reactions occur when weathering and erosion finally reconvert metamorphic rocks to their original sedimentary states. Here again, dissipation of considerable amounts of thermal energy takes place.

But large as volcanic, or diastrophic, or metamorphic energies can be, they all appear relatively small when compared with the general heat flow from the earth. It is well known that temperature almost invariably increases downward in mines, tunnels, and deep wells or bore holes. Thermodynamics thus emphatically requires that heat should flow up toward the surface, the heat flow being simply the product of the temperature gradient times the thermal conductivity. The heat flow will be discussed in more detail in a further section; let it be sufficient to state here that the average value, as measured, is of the order of 1 to 1.5×10^{-6} cal per cm^2 sec. This amounts, for the whole earth, to about 2×10^{20} cal per year, at least 20 times the volcanic energy. This heat flow is, by virtue of its magnitude, the most significant geothermal phenomenon, and is therefore the item that should be accounted for first. Once this is done, other geologic phenomena will probably appear as small perturbations or fluctuations, the explanation of which, we hope, will more or less take care of itself.

Let the reader be reminded that terrestrial heat flow, large as it is, has actually nothing to do with surface temperature, which is almost entirely determined by the amount of heat received from the sun (1.94 cal per cm^2 minute, at normal incidence). The average annual surface temperature of the earth (about 10°C) is that of a black body that would radiate back to space exactly the amount of energy it receives from the sun. This amount being much larger than the terrestrial heat flow, fluctuations in the latter do not appreciably affect the surface temperature, and climatic changes in the past cannot be explained by changes in the earth's output of internal heat.

Internal Structure of the Earth

Before we inquire into the origin of this heat flow, let us recapitulate a few facts concerning the structure of the earth. Most of our information on that score comes from seismology, which studies the manner in which elastic waves generated at the focus of an earthquake propagate between the focus and distant recording stations.

Four clear-cut units have been recognized (Fig. 3); starting from the surface they are, in order of increasing depth. the crust, the mantle, the outer core, and the inner core. The

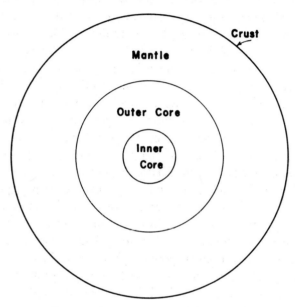

FIG. 3. Internal structure of the earth.

crust is separated from the mantle by the Mohorovičić discontinuity (better known as "Moho"); its thickness, which is about 30–40 km in average continental areas, reflects the surface topography in being somewhat greater under mountain ranges. In oceanic areas, the discontinuity occurs at 10–12 km below sea level, or 5 to 8 km below the oceanic floor. The crust is heterogeneous and consists of sedimentary, igneous, and probably mostly of metamorphic rocks. The most abundant rock exposed at or near the surface of continents is akin to granite, but there is evidence that toward the base of the crust the density increases and the silica content decreases. The oceanic crust consists of a thin veneer of sedimentary muds overlying what are probably basaltic lava flows. The mass of the crust is an almost insignificant fraction of the total mass of the earth.

The mantle extends from "Moho" down to 2,900 km (the radius of the earth is 6,370 km) and accounts for about two-thirds of the earth's mass. It is definitely solid, in the sense that it transmits transverse (or "S") elastic waves which cannot propagate through liquids. Its composition, which must be inferred, is almost certainly close to that of peridotite, a fairly common rock consisting mostly of silicates of magnesium, iron, and calcium, with minor amounts of alumino-silicates of sodium and potassium, and there are good reasons to believe that, in bulk, the mantle cannot be very different from an average stony meteorite. The density increases downward in the mantle somewhat faster than can be expected from self-compression, so that the lower mantle must either contain heavy matter (e.g. iron) or consist of high-pressure polymorphs of silicates of which we have little knowledge; for instance, a high-density polymorph of the mineral olivine, $(FeMg)_2SiO_4$. The outer core is liquid, whereas the inner core is presumably solid; both have relatively high densities (average for both about 10.9 g per cm^3) and are probably metallic, possibly not very different from the metallic meteorites (Fe, with about 10 per cent Ni and other minor elements). It has recently been suggested, for a number of reasons, that the core may also contain about 20 per cent Si[2,3].

The total mass of the earth is known from astronomical and gravity measurements, and its moment of inertia around the polar axis can be calculated from a combination of measurements of the variation of gravity with latitude, geodetic measurements of its flattening, and astronomical determinations of the precession of the equinoxes. This provides information regarding the distribution of mass (density) within the earth. Seismology, on the other hand, supplies (Fig. 4) the velocity of the longitudinal (P) and transverse (S) elastic waves at all depths (the velocity of S is of course zero in the outer core), and these two combined yield the ratio of the

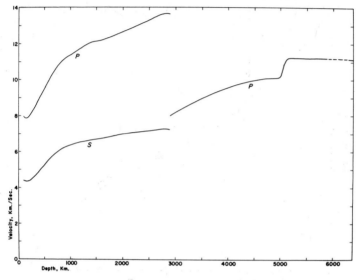

FIG. 4. Velocity of P and S waves in the earth [27, 28].

incompressibility to the density. In a homogeneous medium the increase in density is simply related to the pressure and the incompressibility, the pressure at any depth in turn being a simple function of depth and of the density distribution above that depth. Thus, by a process of numerical integration it is possible to compute possible distributions of density and pressure. Results are not quite unequivocal because there is some uncertainty regarding the exact values of the seismic velocities at depths around 200–400 km, and also because the earth is not homogeneous. We really have more unknown quantities than equations to determine them; consequently, there still remains a certain arbitrariness in our solutions. Nevertheless, we may feel confident that we now do know the density at any depth in the mantle to better than 5 per cent, and the pressure distribution is even better known; densities in the core, particularly the inner core, are more uncertain.

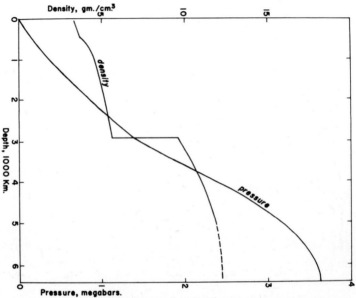

FIG. 5. Pressure and possible density distribution in the earth.

The order of magnitude of the pressures encountered within the earth (Fig. 5) is interesting: about 1.4 megabars (roughly 1.4 million atmospheres) at the core-mantle boundary, and 3.6 megabars at the earth's center. The next problem is to determine what materials could have the observed elastic properties and inferred density at the corresponding pressure. The pressure range unfortunately turns out at the moment to be too high for experimental determinations, and too low to apply rigorous theoretical treatments, such as the Thomas-Fermi theory, which probably become valid only at about 10 or even 100 megabars.

The outer core is, as mentioned, liquid and probably metallic, i.e. a good conductor of electricity. It is of special interest as the source of the earth's magnetic field. The energy to produce and maintain this field is probably thermal (thermally driven convection) so that here again the problem of heat sources and temperature distribution is crucial.

Radiogenic Heat

Prior to the discovery of radioactivity it was generally held that the heat escaping from the earth is the original heat it inherited at birth. According to Laplacian views, the earth was originally part of the sun from which it separated as a very hot gas which later condensed to liquid, with only a thin solid scum ("crust") on its surface. We are no longer quite so sure that the earth ever was very hot—in fact, it may be warming up rather than cooling down—and the original-heat hypothesis was very soon shown to lead to inconsistencies. Kelvin calculated that a body of the size of the earth could not possibly still be losing heat at the present rate if it were more than 100,000,000 years old, and suggested that it might not be much older than 20 or 40 million years. Although there was at that time no precise method of dating rocks, geologists were rightly convinced that the earth is much older than that. The dilemma was solved in 1906 when Rayleigh called attention to the heat liberated by radioactive decay, this heat being essentially equivalent to the kinetic energy of the α and β particles expelled from the disintegrating nuclei. It is now known, for instance, that 1 g of uranium of normal isotopic composition and in equilibrium with all its daughter elements will generate a total of 0.72 cal per year. Thus, the total yearly heat output of the earth (2×10^{20} cal, or 3×10^{-8} cal per g) could be accounted for by an average uranium concentration of the order of 5×10^{-8} g per g. Many rocks actually contain much more uranium than this (10^{-6} g per g).

In addition to uranium, rocks generally contain appreciable amounts of thorium and potassium, and negligible amounts of a few other radioactive elements. Potassium is very abundant, this great abundance compensating in many ways for the relatively small proportion of the radioactive isotope K^{40} and its weak radioactivity. The distribution of potassium in rocks is fairly regular and simple, but uranium behaves rather erratically; different specimens of the same rock type (e.g. granite) may show large fluctuations in their radioactive content. Every progress in analytical methods leads to a lowering of the estimated uranium content; for instance, most recent determinations by neutron activation of radioactive elements in iron meteorites are much lower than previous determinations by chemical methods [4, 5]. Some recent figures are reproduced in Table 1.

TABLE 1. HEAT GENERATION IN ROCKS

Radioactive Content (g/g)

Rock Type	U Total	Th	K Total	Heat Generation (cal/g sec)
Granite	4.0×10^{-6}	1.0×10^{-5}	3.5×10^{-2}	2.2×10^{-13}
Basalt, gabbro	0.6×10^{-6}	3.0×10^{-6}	1.0×10^{-2}	0.3×10^{-13}
Dunite	1.5×10^{-8}	4.6×10^{-8}	1.0×10^{-5}	0.6×10^{-15}
Chondritic meteorite	1.0×10^{-8}	3.5×10^{-8}	8.3×10^{-4}	1.1×10^{-15}
Iron meteorite	1.0×10^{-11}	1.0×10^{-11}	0.3×10^{-6}	0.5×10^{-18}

The main generalization that can be drawn from the data is that, in general, the rate of production of radiogenic heat in rocks varies directly with their silica content, the more siliceous igneous rocks (granites), with about 65 per cent SiO_2, being more radioactive than basalts or gabbros (50–55 per cent SiO_2) which in turn are more radioactive than chondritic meteorites (45 per cent) or dunites (40 per cent SiO_2).

The importance of radiogenic heat is readily seen from these figures. The most common rock in continental areas is

TEMPERATURES WITHIN THE EARTH   49

granite, and the continental crust is about 35 km thick. A column of granite with a cross section of 1 cm² has a volume of 3.5×10^6 cm³ and a mass of 9.4×10^6 g. The total heat generation will therefore be $9.4 \times 10^6 \times 2.2 \times 10^{-13} = 2.1 \times 10^{-6}$ cal per cm² sec. A granitic crust 35 km thick would thus supply more heat than is actually escaping at the surface, even without allowance either for original heat or for radiogenic heat liberated in the underlying mantle and core.

The Equation of Heat Conduction

It now appears that the distribution of radioactive elements in the earth is likely to be the most important factor in determining the total heat generation and the temperature distribution. The relation between them is in theory quite simple, as it is governed by the equation of heat conduction

$$\frac{h}{\rho c} \nabla^2 T = \frac{\partial T}{\partial t} - \frac{\epsilon}{\rho c} \qquad (1)$$

where T is temperature, t is time, ϵ is the rate of heat generation (cal per cm³ sec), h is the thermal conductivity (cal per cm sec deg), c is the specific heat (cal per g deg), ρ is density (g per cm³), and $\nabla^2 T$ is a function of the space coordinates which, for a perfect sphere, reduces to

$$\nabla^2 T = \frac{1}{r^2} \frac{\partial}{\partial r}\left(r^2 \frac{\partial T}{\partial r}\right)$$

r being the distance from any point to the center of the sphere.

Before discussing the general solution of this equation it is profitable to consider some particular cases, notably the steady-state case where the temperature at any depth is assumed to be independent of time $[\partial T/\partial t = 0]$. If ϵ is supposed to be independent of both t and r, solutions are readily found for given boundary conditions (e.g. $T = 0$ at $r = R$, the outer surface of the sphere). If ϵ varies with r, it follows generally that the more the radioactive material is concentrated toward the surface, the lower the temperature will be at any depth below the surface. If, for instance, all heat sources were concentrated at the very center of the sphere, the temperature would increase linearly from the surface downward at the same rate at which it increases near the surface; the temperature at the center of the earth on this model would be $30 \times 6{,}370 = 191{,}000°C$. If, on the other hand, all heat sources were spread on an infinitesimally thin layer on the surface at temperature 0° the temperature at the center would also be 0°. The rule of thumb is that temperature increases downward until the heat sources are reached; wherever a heat source is gone through, the rate of temperature increase (temperature gradient) decreases proportionally to the intensity of that source.

The steady-state case is, of course, only an approximation. The half-life of the main radioactive elements is commensurate with the age of the earth, which means that the amount of radioactive materials present has changed appreciably during geologic times; thus ϵ in equation (1) is a function of both r and t. A large amount of effort has been spent on computing solutions of (1) under various assumptions, with a few significant results. In the first place, ϵ is an unknown function of r; we do *not* know a priori how the radioactive sources are distributed, and most simple models that can be devised ultimately turn out to be wrong. Furthermore the density, ρ, is a function of depth and of T, because of thermal expansion. This, however, is only a minor factor compared with the uncertainties regarding h, the thermal conductivity. Thermal conductivity depends on composition, being different for different types of rocks. It depends on pressure in an almost unpredictable way, and it also depends on temperature. In fact, the main mechanism of conduction in the deep earth is not known. Although one generally assumes that phonon (elastic wave) conductivity is dominant, it is at least possible that radiation transfer could become more effective at depths where the temperature exceeds 1,500 or 2,000° K [6, 7]. The radiation conductivity is very senstive to T (it varies as T^3) so that it is not permissible to treat h as a constant. It now appears that, instead of using the heat-conduction equation to determine T, as has been generally done, it will be necessary to determine T by some other method, and use the equation to determine ϵ and h.

Surface Heat Flow

As mentioned above, a very simple calculation suffices to show that if the continental crust consisted entirely of granite, its heat output should be greater than has been observed. It follows that continents must also contain rocks less radioactive than granites, a conclusion to which geologists would gladly adhere. Nevertheless, continents *do* contain a large amount of granitic rocks which, unless our analytical data are all wrong, must contribute a notable fraction of the continental heat flow. Granites are typically absent in oceanic areas; one would expect, therefore, the heat flow to be notably less on the floor of the oceans than on continents.

The determination of the heat flow q is, in principle, very simple: one has only to measure the temperature gradient (dT/dn, where n is depth), and the thermal conductivity h of the rocks in which the gradient is measured; then $q = h(dT/dn)$. One could, for instance, drill a hole, measure h on the core samples and measure T at different depths in the hole. One must, of course, ascertain that the temperature has not been disturbed by drilling, or by fluid circulation in the bore hole. It also turns out that the thermal conductivity of rocks is so small that it takes a very long time to eliminate effects arising from climatic variations at the surface; if one measures, for instance, the heat flow in an area which has been recently (geologically speaking!) subjected to glaciation, the measured thermal gradient will reflect, down to depths of a few hundred feet, the changes in surface temperature since removal of the ice. If measurements are carried out in mountain ranges, it becomes important to know for what length of time the surface has stood at its present elevation, as the mean annual surface temperature depends on altitude. Allowance must thus be made for what is thought to be the thermal history of the area; as this is uncertain, heat-flow determinations are affected by a relatively large uncertainty. Nevertheless, measurements in continental areas show a remarkable consistency. They fall in the range 0.7 to 3.0 microcalories per cm^2 sec, the higher values such as that (2.9) found by Misener [8] at Resolute Bay being accounted for by the proximity of the ocean and recent changes in shore-line configuration [9]. The average value for continents stands at about 1.2.

Oceanic areas. We have just seen that a lower value should be expected on the ocean floor. A large number of heat-flow values has recently been obtained from the Atlantic and in greater numbers from the Pacific [10, 11]. The average, if anything is *higher* than for continents; for instance, a recent set of 36 measurements from the southeastern Pacific [11] has a mean value of about 1.8. Averages, however, are not very meaningful, as the heat flow shows considerable local variation; some stations yield very high values, while others are very low (maximum 8.09; minimum 0.14). Nothing like this has been observed on land. There is a broad relationship to submarine topography, in that high heat-flow values are commonly found on rises, while the very low values are generally characteristic of deep trenches (Fig. 6).

These observations are quite unexpected and very puzzling. Birch [12] has pointed out that it is perhaps not surprising that the oceanic heat flow should be about the same as the continental one, as the total heat output one could expect from a mantle with the radioactive content characteristic of chondritic meteorites would be just about 1 microcalorie per cm^2 sec. The only difference between oceans and continents

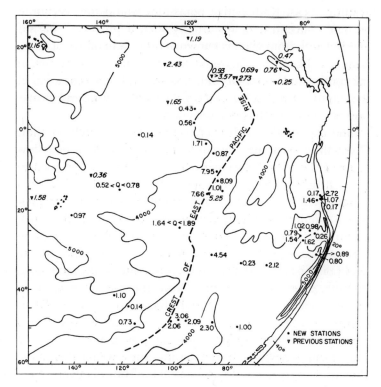

Fig. 6. Heat flow in the eastern Pacific [11].

would be that the latter represent areas where the original radioactive matter of the mantle, or a large part of it, has been swept up toward the surface and concentrated in the crust [10]. This is, in fact, an attractive suggestion. Geologists do not know how to account for the existence of continents, and it seems as likely as not that they have indeed grown piecemeal by accretion of molten material (magma) rising from the mantle. The melting relationships in the multicomponent system represented by a stony meteorite (if that be the composition of the mantle) are such that the first liquid to form would be considerably enriched in alumino-silicates of sodium and potassium, indeed the main ingredients of granitic rocks. Radioactive potassium would thus certainly become concentrated upward by such a process of partial melting and, from what is known of the geochemistry of uranium, it is likely that it would also become concentrated in the continental rocks. Oceanic areas would represent areas of the mantle in which the process of partial melting and formation of continents has not yet occurred.

The explanation, however, cannot be so simple as that. Given the radioactive content of the mantle and the heat flow at its surface, the heat conduction equation can be used to give a steady-state solution, using the normal value of thermal conductivity as measured on peridotites or meteorite samples, which should surely be approximately correct near the top of the mantle. When this is done, it turns out that the temperature increases so rapidly with depth that it is impossible to avoid melting even when allowance is made for the increase in melting point with pressure. Now the mantle cannot be wholly molten, for its essential characteristic is indeed its ability to transmit S waves, which characteristically do not propagate through liquids. To be sure, some molten material must form somewhere in the mantle, since we see it rising to the surface as magma, for instance in the Hawaiian volcanoes; but volcanism is an intermittent phenomenon which throughout geologic time has been restricted, at any one time, to fairly small and scattered areas. The total amount of molten stuff at any time must be relatively small, judged also from the earth's resistance to tidal deformation; formation of liquid as evidenced by volcanic action appears more in the nature of a local and transient fluctuation.

A further point regarding the heat flow from the mantle is that the latter is so big, and its normal (phonon) conductivity so small even after allowance for possible effects of pressure, that no heat generated in its deeper parts could reach its surface in times commensurate with the age of the earth. Slichter [13] called attention to this fact many years ago: heat reaching the surface today cannot have traveled more than a few hundred kilometers. Thus, the oceanic heat flow would correspond to the radioactivity of the upper mantle, not to that of the whole mantle.

This last statement applies, of course, only if we consider the phonon conductivity. If heat transfer occurs mainly by radiation, with an effectively much larger conductivity, more heat could be carried faster with a lower average gradient. But radiative transfer can be appreciable, as stated above, only when the temperature is about that at which melting would occur at normal pressures, so that it cannot be effective in the topmost 200 kilometers or so of the mantle. Thus, if the oceanic heat flow measurements are correct, melting in the uppermost mantle can be avoided only by a strong upward concentration of radioactivity, as occurs in the continents. How this concentration could have been effected is not known.

Regional variations in heat flow on the Pacific floor are difficult to understand on any conduction mechanism. Seismic studies do not reveal notable differences in composition of the mantle between areas of high and low heat flow; there is no basis whatever, except the heat flow measurements themselves, for postulating local differences in radioactivity of the upper mantle beneath rises or deeps. It has been suggested that the regional pattern indicates transport of heat by convection rather than conduction, high heat-flow values corresponding to currents of hot material rising to the surface, while low heat values occur where cold material is descending.

Convection

The suggestion that convection does occur in the mantle is not a new one; in fact it antedates any measurements of the oceanic heat flow. Broadly speaking, convection, or mass transport, is likely to occur in any fluid in which there exist horizontal temperature gradients, or vertical temperature gradients exceeding the "adiabatic" value, as these temperature gradients will produce density gradients and disturb any pre-existing hydrostatic equilibrium. To be sure, the mantle is not a fluid in the real sense; yet it is capable of plastic deformation, the only difference between the rheological properties of the mantle and those of an ordinary fluid being that the former has presumably *1*) a yield point (i.e. a minimum stress below which deformation, if any, is purely elastic), and *2*) a very high effective "viscosity," the latter depending presumably on the stresses. For instance, it has been calculated that a Newtonian liquid that would deform at the rate at which the mantle deforms when local loads, such as ice-cap, are applied on the earth's surface must have a viscosity of the order of 10^{21}–10^{22} cgs. But having either a yield point or a very high "viscosity" need preclude the possibility of convection in systems with sufficiently large linear dimensions, as dimensional analysis readily shows. Thermal stresses that would obviously not induce convection in a block of stone can certainly do so if the dimensions of the block are of the order of 10^8 cm.

Convection can be conveniently considered by adding to the right-hand side of equation (1) a term $\bar{v} \cdot \text{grad } T$, where $\bar{v}$ is the velocity of deformation. Comparison of orders of magnitude lead Slichter [13] to the conclusion that the rate of heat transport by convection would greatly exceed that by conduction even if velocities are as low as 0.01 cm per year. For the sake of comparison we may note that deformation at the rate of a few centimeters per year *has* been observed on the surface of the earth, notably on either side of the San Andreas fault in California.

Temperatures Inferred from Solid-state Theory

The reader must be well convinced by now that the distribution of radioactive elements in the earth and the mechanism of heat conduction are of utmost importance to its heat economy. Yet there is no way of determining directly what they are and whether the earth is at present cooling or warming; i.e. losing more or less heat than is actually generated in it. Clearly, the heat balance cannot be set up from the surface heat flow alone, and one must turn to other methods. It is interesting to see whether any other measurable properties of the earth can give us an indication of its internal temperature.

Theoretically, the problem should be soluble, thanks to seismology, if the internal composition were exactly known. For the velocity of elastic waves in a substance of given composition is a function of pressure and temperature. We know the velocity of both P and S waves at most depths in the mantle. The pressure at any depth turns out, by a happy coincidence, to be nearly independent of composition (provided that mass is distributed so as to give the correct moment of inertia) and thus fairly accurately known. It would then be sufficient *1*) to calculate what the velocities should be at a given pressure and zero temperature, by solid-state theory; *2*) to compare with observed velocities; *3*) to assign the difference between calculated and observed velocities to the effect of temperature; and *4*) to compute the temperature therefrom, again from theory. Needless to say that solid-state theory is still unable, in the absence of experiments, to cope satisfactorily with steps *1* or *4*, and the trouble with experiments is, quite apart from the pressure range involved, that we do not know what to experiment on, as the chemical composition and physical state are not known with certainty. The problem has been reviewed in detail by Birch [14] who shows how the problem could be solved, given certain coefficients such as the pressure and temperature derivatives of the isothermal compressibility. At present one can only guess what the order of magnitude of these quantities will be, and make a rough estimate of the temperature, which turns out to be of the order of a few thousand degrees, with 10,000° as an uppermost limit. The difficulty, fundamentally, is that in the earth the effect of temperature on elastic properties, such as the velocity of seismic waves, is apparently much less than the effect of pressure; thus, temperature appears as a second-order effect which cannot be separated from the much larger effect of pressure until the latter is more accurately known. One may be hopeful that progress in experimental high-pressure techniques will soon yield much-needed information.

In the meantime, attempts have been made to establish upper and lower limits to the temperature prevailing at various levels, particularly at the mantle-core and outer-inner core boundaries.

Melting relationships. As it is generally held that the temperature everywhere in the mantle—except at volcanic foci—is less than the melting point, a determination of the melting point as a function of depth would also serve to establish an upper limit to the temperature.

Properly speaking, a rock, being a multicomponent system, does not have a definite melting point; there is a temperature range in which melting occurs, generally accompanied by a complicated succession of reactions between the liquid and remaining solid phases (incongruent melting, and so forth). The temperatures at which melting begins and ends depend very much on composition. However, the initial slope of the melting-point curve has been determined for a number of silicate minerals and is generally found to be between 5 and 20° per thousand bars. The Clausius-Clapeyron relation

$$\frac{dT_m}{dP} = \frac{\Delta V}{\Delta S} \qquad (2)$$

(where ΔV and ΔS are, respectively, the volume and entropy of melting) leads to the prediction that dT_m/dP will in general decrease at higher pressure, as ΔV certainly does (the liquid phase is more compressible than the solid one), while ΔS contains a term proportional to ΔV and a purely configurational term that is not likely to be affected. The slope of the melting-point curve also seems to be inversely related to the melting point itself, so that pressure affects least the minerals that melt at the highest temperature. Thus, forsterite (Mg_2SiO_4), which is closely related to olivine [$(FeMg)_2SiO_4$)], an important constituent of peridotites and stony meteorites, is likely to be present in the mantle, and probably has an average melting-point gradient not greater than 3.5° per 1,000 bars; its melting point at the bottom of the mantle (1.4 × 10^6 bars) is thus not likely to be more than 5,000° higher than its normal melting point which is 1,890° for pure Mg_2SiO_4. (The melting range of a common olivine with an Mg/Fe ratio of 5 is 1,750°–1,850°C; the possibility that the Mg/Fe ratio decreases with depth in the mantle introduces a further complication that must be ignored here.) Olivine is one of the last minerals to melt at normal pressure, so that the temperature at which melting would begin at the bottom of the mantle would be less than about 7,000°C. This could be taken as an upper limit to the temperature at the bottom of the mantle.

Similar results were obtained by Uffen [15], using Lindemann's theory of melting. On this theory, melting occurs when the amplitude of the thermal oscillations in a solid exceeds a certain value which depends on the atomic spacing. Now the amplitude is related to the thermal energy of oscillation which can be determined at a given temperature if a certain parameter θ, the "Debye temperature" of the solid, is known. The Debye temperature plays an important role in all theories of the earth's interior. It is defined formally as $\theta = (h\nu_m/k)$ where ν_m is the highest vibration frequency of which the lattice is capable, while h and k are, respectively, the Planck and Boltzmann constants. The maximum frequency ν_m depends on the lattice spacing (density) and interatomic forces such as determine the incompressibility; thus, ν_m must be related to the velocity of elastic waves, which also depend on these two quantities, It is possible, using Debye's theory, to find how θ varies with depth in the mantle and, correspondingly, how the melting point should vary. By this method Uffen finds that the melting point at the core boundary should be less than three times what it is at 100 km depth; this again sets an upper limit of about 5,000°C.

Neither these nor the preceding calculations take into account the possibility of phase changes. As indicated earlier it seems an unavoidable conclusion that the mantle is not completely homogeneous; in particular, the lower part of it must have a higher density than would result purely from compression; that is, it must contain either denser material or denser polymorphs of the same materials. It is impossible to predict what effect this would have on the temperature at which melting begins.

Mantle-core boundary. Since the outer core is liquid, the possibility exists of bracketing the temperature at the mantle-core boundary, as it must be less than the melting point of mantle material but higher than the melting point of core material. Whatever the composition of the core may be, there are good reasons for believing that it consists dominantly of iron, with probably some nickel (the average iron meteorite contains 10 per cent Ni), and about 20 per cent silicon. It would therefore be of considerable interest to determine the melting point of iron at pressures on the order of those prevailing in the core.

Much thought has been devoted to this problem. In 1953, Simon [16] proposed to apply an empirical equation that he had found suitable for describing his experimental results on the variation with pressure of the melting point of a number of easily fusible substances

$$\frac{P}{a} = \left(\frac{T}{T_0}\right)^c - 1 \qquad (3)$$

where T is the melting point at pressure P, T_0 is the melting point at zero pressure, a and c are empirical constants. The constant a has the dimensions of pressure and is interpreted by Simon to be the "internal pressure $(\partial U/\partial V)_T$," where U is internal energy, in the usual thermodynamic sense. The product ac may be found if the initial slope of the melting-point curve at $P = 0$ is known. Simon suggested that, by analogy with other substances, a for iron should be taken as 150,000 atm and $c = 4$. Gilvarry [17] showed that Simon's equation follows directly from a slightly modified form of Lindemann's law, and he also showed that c should be related to γ, Grüneisen's constant, as

$$c = \frac{6\gamma + 1}{6\gamma - 2} \qquad (4)$$

where γ is to be evaluated at the melting point at zero pressure.

Grüneisen's constant, γ, plays an important role in solid-state theories of the earth's interior; it is defined, in Debye's theory, as $\gamma = -(d \ln \theta / d \ln V)$, where θ is, as before, the Debye temperature of the solid and V its volume. Simple transformations lead to the equivalent relation

$$\gamma = \frac{\alpha K_s}{\rho c_p} = \frac{\alpha K_t}{\rho c_v} \qquad (5)$$

where α is thermal expansion; K_s and K_t are, respectively, the adiabatic and isothermal incompressibilities; ρ is density; and c_p and c_v are, respectively, the specific heat at constant pressure and constant volume. Grüneisen's constant has the property that it has nearly the same value for most substances, the common range being about 0.8 to 2.0. Theoretically it should depend on volume (thus also on pressure and temperature), but the variations are slight. Gilvarry's own evaluation of γ led him to adopt $c = 1.9$ for iron, with $a = 347,000$. Strong [18] has recently determined experimentally the melting point of iron up to 96,000 atm and finds that it fits the Simon equation with $c = 8$, $a = 75,000$. The melting point of iron, extrapolated to pressures such as exist in the core by

Simon, Gilvarry, and Strong is shown in Figure 7. The values at the core boundary ($P = 1.4 \times 10^6$ atm) are: Simon, 3,250°C; Gilvarry, 3,900°C; Strong, 2,340 ± 50°C. Strong's value should be preferred, as it is more tightly related to experimental data. However, the value of γ corresponding to $c = 8$ is 0.4, which seems low; a slightly different fit of

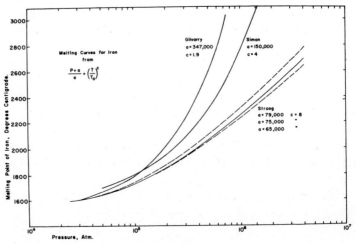

FIG. 7. The melting point of iron, as a function of pressure [18].

Simon's equation to the experimental data would allow a smaller value of c and a slightly higher melting point.

The inner core-outer core boundary. Seismological observations strongly suggest that the inner core is solid. Its density is only approximately known, for the reason that its radius (1,200 km) is relatively small and therefore it contributes very little to the total mass of the earth and still less to its moment of inertia. It has been proposed that it has the same composition as the outer core, which would put the boundary of the two exactly at the melting point. The pressure there is about 3 million atmospheres; the melting point of iron as extrapolated by Simon is close to 3,800°C while the Strong extrapolation gives 2,600 ± 200°C. Gilvarry's corresponding value is about 5,900°C.

These melting temperatures are not meant necessarily to represent the actual temperature. Whatever the core consists of, it is surely not pure iron, and the effect of adding Ni, Si, and so on to iron would be to lower the melting point; at ordinary pressure, addition of 20 per cent Si to iron lowers its melting point by 300°. What these additions would do to the slope of the melting-point curve is not known. Besides, the Lindemann-Simon-Gilvarry theories of melting have, in common, the awkward feature that melting is predicted from the properties of the solid only. Now melting corresponds to equilibrium between solid *and* liquid, and one would expect the properties of the liquid to enter somehow into the relation of melting point to pressure. Determinations of melting point, as applied to the earth's interior, retain considerable uncertainty.

The Adiabatic Gradient

We have seen that the melting-point curve for the mantle sets an upper limit to its temperature. The adiabatic gradient on the other hand, is likely to set a lower limit. The adiabatic gradient has the interesting property of being the minimum temperature gradient necessary to start convection in a fluid; it is also the gradient that would prevail purely as a result of self-compression. If a substance is compressed adiabatically, that is, without exchanging heat with the surroundings, the pressure coefficient of the temperature is found by simple thermodynamics to be $(\partial T/\partial P)_S = (\alpha T/\rho c_p)$, where all symbols have the same significance as before. In a body in hydrostatic equilibrium, such as the earth, $dP = -g\rho\,dr$, where g is the acceleration of gravity. Thus, under adiabatic conditions,

$$\frac{1}{g}\frac{d \ln T}{dr} = -\frac{\alpha}{c_p} \qquad (6)$$

If the earth had grown by accretion and without heat sources of any kind in it, the temperature inside would be found, given the surface temperature, by a downward integration of this relation. If the earth contains heat sources and is capable of convective motion, convection would tend to transport heat at a rate sufficient to bring the temperature gradient down to the adiabatic value, at which convection would stop. Thus, the adiabatic gradient represents the minimum gradient that is likely to prevail.

With some assumptions and simplifications, the adiabatic gradient may be estimated from seismic data [19]. The velocities of the P and S waves in a solid are such that

$$v_P^2 - \frac{4}{3} v_S^2 = \frac{K_s}{\rho} \tag{7}$$

where K_s is, as before, the adiabatic incompressibility and ρ is density. Call this ratio φ; i.e. $v_P^2 - \frac{4}{3} v_S^2 = \varphi$. Now, by definition of the Grüneisen constant, γ,

$$\gamma = - \frac{d \ln \theta}{d \ln V} = \frac{\rho}{\theta} \frac{d\theta}{d\rho} = \frac{\alpha K_s}{\rho c_p} = \frac{\alpha \varphi}{c_p} \tag{8}$$

In Debye's theory, $\theta = h\nu_m/k$, where

$$\nu_m = \bar{v} \left(\frac{3N}{4\pi}\right)^{2/3} \tag{9}$$

where $\bar{v}$ is an average velocity such that $3\bar{v}^{-3} = v_P^{-3} + 2v_S^{-3}$. N is the number of particles per unit volume and can be estimated, given the density and approximate chemical composition. Thus, given possible values of the density at all depths in the mantle, and the seismic velocities from which φ may be calculated, one can determine ν_m at each depth. Having ν_m, simple arithmetic gives θ, and a plot of θ versus ρ gives $d\theta/d\rho$, from which γ may be computed. Then $\alpha/c_p = \gamma/\varphi$ can also be found. The acceleration of gravity g depends, of course, on position within the earth, but may be found without difficulty, given a distribution of density compatible with the earth's mass and moment of inertia; rather fortunately it turns out that, due to the high average density of the core, g remains very nearly constant in the mantle where it varies from 980 cm sec^{-2} at the surface, to 1,040 cm per sec^2 at the core boundary. Thus by integrating (6) from the Mohorovičić discontinuity downward, the adiabatic temperature may be found at all depths in the mantle. The temperature at the Moho discontinuity is, of course, not precisely known; it is probably of the order of 150 or 200°C in oceanic areas, and estimates for continental areas, using the heat conduction equation, yield values between 300 and 700°, depending on how much radioactivity is assumed to be present in the crust. Because of some uncertainties regarding the exact value of the seismic velocities in the uppermost mantle, it is actually preferable to integrate from 200 km downward; if T_{200} represents the actual temperature at that level, the adiabatic temperature at the core boundary is found to be about 1.5 T_{200}, and the average adiabatic gradient in the mantle is 1.7×10^{-4} T_{200}. For $T_{200} = 1,500°K$ the adiabatic temperature at the core boundary is $\sim 2,250°K$, and the average adiabatic gradient is about 0.28° per km.

Thermal Expansion

There is, of course, no a priori reason why the temperature in the mantle should approach the adiabatic value, but there may be some indications that it does. We have just seen how the quantity α/c_p may be determined, under proper assumptions, from seismic data. Now c_p, the specific heat at constant pressure, is a quantity that should theoretically be rather insensitive to either P or T; from thermodynamics we have

$$\frac{1}{c_p}\left(\frac{\partial c_p}{\partial T}\right)_P = \frac{\alpha \gamma}{1 + T\alpha\gamma}$$

which is a small quantity of the order of α, while

$$\left(\frac{\partial c_p}{\partial P}\right)_T = -T\left(\frac{\partial^2 V}{\partial T^2}\right)_P$$

which is also small. The specific heat at high temperature (that is, above the Debye temperature) is also insensitive to chemical composition, as theory predicts. Thus, c_p can be estimated without danger of gross error and, therefore, α itself can be found. Its value in the mantle near the core boundary is about $(1.0 \pm 0.2) \times 10^{-5}$ per deg. As the temperature at the core boundary is likely to be between 2,000° and 5,000°K, αT is likely to be between 1.6×10^{-2} and 6×10^{-2}. Incidentally, θ is close to 1,300°K so that T/θ lies in the range 1.5–4.

It has been noted [20] that the thermal expansion of a large number of solids follows a law of the type

$$\alpha T = b(T/\theta)^{1.5} \qquad (10)$$

where b is a constant which depends only on pressure and which has very nearly the same value $(24.7 \pm 4.5) \times 10^{-3}$ for a large number of substances (e.g. diamond, NaCl, SiO_2, many silicates and oxides, and several metals). The law is valid at least from $\alpha T = 1.3 \times 10^{-3}$ to $\alpha T = 0.2$, and from $T/\theta = 0.14$ to $T/\theta = 4.3$. It is therefore also likely to be valid, *without extrapolation*, at the core boundary. The only difficulty in applying this relation to find T, α and θ being known, is to estimate the pressure dependence of b. From thermodynamics it is found that

$$\frac{K}{b}\frac{db}{dP} = \frac{1}{\alpha K}\left(\frac{\partial K}{\partial T}\right)_P + \frac{3}{2}\gamma \qquad (11)$$

where K is the isothermal incompressibility. Fortunately, it is possible to set an upper limit to the right-hand side of (11), from which a minimum value of b at the core boundary can be found. This in turn leads to a maximum temperature of 2,700°K. However, θ is involved as a third power, and θ

Temperatures in the Mantle

We may now attempt a rough synthesis of our knowledge concerning the mantle. We have just found that the temperature at the boundary of the core is likely to be less than 2,700°K, but it must be more than the melting point of whatever forms the outer core (2,600°K for pure Fe, but probably less for the actual composition). It is perhaps not an accident that the two figures should overlap so closely, for the lower mantle might contain a notable proportion of metallic phase (most stony meteorites do), so that the mantle-core, or solid-liquid, boundary should occur precisely where melting of the metallic phase begins. It would seem, at any rate, that a temperature of 2,600°K might be right, give or take 200 or 300°. As the gradient in the mantle is not likely to be less than adiabatic, the ratio $T_{2900}/T_{200} = 1.5$ sets an *upper* limit of 1,700°K, or roughly 1,500°C, for the temperature at 200 km. Only if radiative transfer of heat is even more effective than convection could the temperature at 200 km be much higher than this, and radiative transfer is *not* likely to be very high at the relatively low temperatures that seem to emerge from our calculations (remember that h_{rad} is proportional to T^3). On the other hand the temperature at 200 km cannot be much less than 1,500°C, as the average heat flow in oceanic areas implies a gradient of at least 10 to 15° per km at the top of the mantle. Temperature must increase rapidly from Moho downward, reaching possibly 1,100–12,00°C near 100 km depth, and then flattening out to nearly the adiabatic value (Fig. 8). The rather sudden change in slope of the

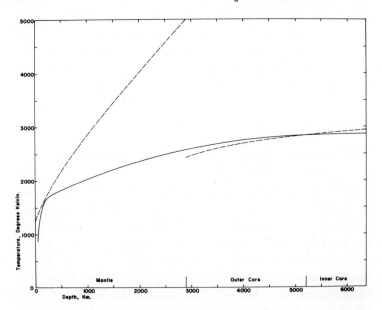

Fig. 8. Possible temperature distribution in the earth. The dashed line tentatively represents the melting point.

temperature-depth curve may be due to a high concentration of radioactivity in the uppermost mantle, but is more likely to indicate a transition from conduction (phonon) to convection. The rapid rise in temperature in the upper 200 km is probably the cause of the relatively low seismic velocities observed around 100 km depth, as temperature tends to decrease the velocity of elastic waves.

Volcanoes in oceanic areas invariably produce basaltic lavas, of a composition such as could result from partial melting of meteoritic material. The temperature at which melting would start, and basaltic magma would begin to form, is around 900–1,000° at ordinary pressure, and rises initially at the rate of about 10° per 1,000 bars; at a depth of 200 km, melting would probably begin at 1,500 or 1,600°C. There are reasons to believe that basaltic magma forms, when it does, in the uppermost 200 km. Between 100 and 200 km depth, the melting point comes sufficiently close to the deduced temperature to show that relatively minor disturbances could lead to melting (Fig. 8); for instance, melting would almost surely occur if convective matter, leaving the core at the assumed temperature of 2,600°K, were to rise within 100 km or so of the surface.

The Outer Core

It is particularly interesting to consider the outer core, because that is where the earth's magnetic field originates. The outer core is believed to be in convective motion, the evidence being partly that convection is necessary to explain the magnetic field, but also that changes on the rate of rotation have been observed that cannot be accounted for otherwise [21]. One feels confident that the temperature gradient in the core must be higher than the adiabatic value; how much higher we do not know. Viscosity, which is very low, would have a negligible effect on convection because of the large linear dimensions of the system, but the effects of the internal magnetic field on inhibiting convection are not precisely known, although they are probably small. If we make, as before, the assumption that the inner core is solid, with the same composition as the outer core, the temperature gradient must also be less than the melting-point gradient.

The adiabatic gradient β may be estimated as before:

$$\beta = \frac{\alpha}{c_p} g T$$

and $\alpha/c_p = \gamma/\varphi$, by (8). At the top of the core $\varphi = V_P{}^2 = 0.66 \times 10^{12}$ cm^2 sec^{-2}. For γ let us use the value 0.4 corresponding to Strong's experiments on the melting point; this leads to $\alpha/c_p = 0.6 \times 10^{-12}$. The specific heat of liquid iron is 0.18 cal per g deg at the melting point at ordinary pressure, and is not likely to vary much along the melting curve. Then

$\alpha = 0.45 \times 10^{-6}$, which seems low but which is exactly the value obtained by Jacobs [22] by an entirely different method based on a hypothetical relation between thermal expansion and compressibility. The adiabatic gradient β then comes out as 0.15° per km for $T = 2,600°K$; it would probably decrease with depth, as g decreases with increasing proximity to the center of the earth, while φ also increases.

Let us now consider the energy balance of the core, which fortunately can be assessed by two somewhat unrelated methods. If the core is convecting, as we believe it is, more heat must be transferred per unit time than could be disposed of by conduction under a gradient less than adiabatic. To evaluate the latter we need to know the thermal conductivity h. For metals, the electronic (and probably dominant) part of the thermoconductivity is related to the electrical conductivity σ by the Wiedemann-Franz relation which makes the ratio h/σ directly proportional to T. The electrical conductivity can be estimated from the rate at which the magnetic field changes on the surface of the earth. Bullard [23] uses a value $\sigma = 3 \times 10^{-6}$ electromagnetic units and shows, by an analysis of the rate of westward drift of the magnetic field, that this figure is unlikely to be wrong by a factor of more than 2; various extrapolations of σ for liquid iron, under conditions prevailing in the core, range from 1 to 5×10^{-6}. Using again $T = 2,600°K$, gives $h = 0.045$ cal per cm sec deg. To this we add a small contribution for the phonon conductivity. The total is not likely to be more than 0.08. The total heat flow for an adiabatic gradient of 0.15° per km is then equal to 0.12×10^{-6} cal per cm² sec, or 1.8×10^{11} cal per sec for the whole core. If we use Gilvarry's figures we obtain a figure about one order of magnitude higher.

The heat output of the core can be estimated in a different way. The magnetic field of the earth is produced ultimately by electric currents generated in the core, as in a dynamo, by the displacement (convection) of conducting material that moves across lines of magnetic force within the core. Because the core is not a perfect conductor, the electrical currents tend to decay, the time of decay being a simple function of the size of the core and of its conductivity. But if the currents decay, so does the magnetic field H, to which an energy $H^2/8\pi$ per unit volume is associated. To maintain the field thus requires an expenditure of energy E which is variously estimated as 0.5×10^9 cal per sec (Hide [24]) and 2×10^{10} cal per sec (Bullard [23]), the difference arising mainly from differences in the estimated intensity of the magnetic field within the core. Now, the energy comes essentially from mechanical energy of motion of the fluid. If convection in the core operates between temperatures T_1 and T_2 where T_1 and T_2 could be identified, respectively, with the temperatures at the bottom and top of the convecting units, the maximum thermodynamic efficiency, which is the ratio of mechanical work to heat output, would be $(T_1 - T_2)/T_1$. Thus, the total heat output must be at least $ET_1/(T_1 - T_2) \approx 10E \approx 2 \times 10^{11}$ cal per sec, if we use Bullard's figures. This agrees well with the estimate of the minimum heat flow required to maintain convection.

When the dynamo theory of the magnetic field was first discussed, about ten years ago, there seemed to be no difficulty in accounting for this heat output, as the radioactivity of iron meteorites was believed at that time to be equivalent to 10^{-10} cal per g sec, or roughly 2×10^{11} cal per sec for the whole core. As shown in Table 1, recent determinations yield much lower figures—about 0.5×10^{-18} cal per g sec, or 1×10^9 cal per sec for the whole core. This would certainly be insufficient to maintain convection and a magnetic field. A search for other possible sources of heat in the core seems to lead only to one: the latent heat of crystallization of the inner core, which must, on this argument, be growing. A rough calculation is as follows [25]. The latent heat of melting of pure iron at ordinary pressure is 67 cal per g, and the corresponding entropy of melting is 2 cal per mole deg. This is very close to the "communal" entropy, a purely configurational term that is not likely to vary with pressure and temperature, so that one might hazard the guess that the entropy of melting of iron in the core is still close to 2 cal per mole deg. As the melting point at the pressure corresponding to the inner core boundary is probably less than twice that at ordinary pressure, the heat of melting is less than twice 67 cal per g, say 120 cal per g. To supply at least 2×10^{11} cal per sec thus requires crystallization of about 1.6×10^9 g per second. The mass of the inner core is about 1×10^{26} g, so that its age could not be much greater than 2 billion (2×10^9) years; a rough estimate is about 3 billion years [25]. The present rate of cooling is of the order of 10–50 deg per billion years.

Thermal History of the Earth

Let us now return to the old and much debated question as to whether the earth is slowly cooling or slowly warming. Most recently, Lubimova [25] has attempted to show, by use of the heat conduction equation, that there must be a slight cooling of the outer layers with simultaneous heating of the interior (Fig. 9).

It would now seem that Lubimova's conclusions cannot be strictly adhered to. They are based essentially on *1*) a very low "initial" temperature of formation, *2*) an amount of radioactivity in the core much higher than what recent analytical data suggest, and *3*) a hypothetical event occurring three billion years ago and resulting in the formation of continents. The evolution of the core is particularly interesting. According to Lubimova, who uses Gilvarry's melting point for iron, melting began about 1.2 billion years ago. This is certainly wrong, for there is paleomagnetic evidence (from the magnetization of old rocks) that the earth's field has existed for at least 2 billion years. If we use Strong's results for

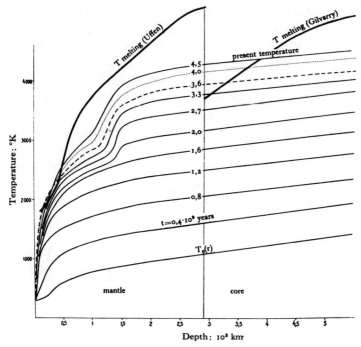

FIG. 9. Thermal history of the earth, according to Lubimova.[26]

the melting point and a correspondingly low temperature of the core, it would now be entirely liquid, which is again wrong. Lubimova's calculations neglect the latent heat of melting, which is serious. On the whole, and unless our data on the composition of the core and its radioactivity are completely wrong, the core must be cooling, not heating. This, in turn, implies a much higher initial temperature than that assumed by Lubimova.

A high initial temperature is, of course, not inconsistent with the view now generally held by geochemists that the earth formed by accretion of dust and gas initially at very low temperature, of the order of a few degrees absolute. The total gravitational energy released by condensation of an object of mass M and radius R is $3GM^2/5R$, where G is the gravitational constant; this energy is equivalent for the earth to 9,000 cal per g, which would be sufficient to vaporize the whole of it. The actual temperature prevailing during formation could be quite high, and would depend essentially on the rate of accretion and rate at which heat can be radiated away. Initial temperatures above the melting point of iron are required in Ringwood's [2] theory of formation of the metallic core by chemical reduction of the original meteoritic matter, which is, to the present writer at least, the only sensible theory that has been presented to account for the presence of a core.

REFERENCES

A GENERAL ACCOUNT of the physics of the earth may be found in: J. A. Jacobs, R. D. Russell, and J. Tuzo Wilson, *Physics and Geology.* New York, McGraw-Hill (1959). A more extensive bibliography appears in: J. Verhoogen, Temperatures within the earth, pp. 17–43 in *Physics and Chemistry of the Earth*, vol. 1. London, Pergamon Press (1956).

1. GOGUEL, J. Introduction à l'étude mécanique des déformations de l'écorce terrestre. *Mem. Carte Geol. France* (1948).
2. RINGWOOD, A. E. *Geochim. et Cosmochim. Acta*, **15,** 257 (1959).
3. MACDONALD, G. J. F., and L. KNOPOFF. *Geophys. J.*, **1,** 284 (1958).
4. REED, G. W. in *Researches in Geochemistry*, P. H. Abelson, Ed., pp. 458–473. New York, Wiley (1959).
5. STOENNER, R. W., and J. ZÄHRINGER. *Geochim. et Cosmochim. Acta*, **15,** 40 (1958).
6. CLARK, S. P. *Trans. Am. Geophys. Union*, **38,** 931 (1957).
7. LAWSON, A. W., and J. C. JAMIESON. *J. Geol.*, **66,** 540 (1958).
8. MISENER, A. D. *Trans. Am. Geophys. Union*, **36,** 1055 (1955).
9. LACHENBRUCH, A. H. *Bull. Geol. Soc. Amer.*, **68,** 1515 (1957).
10. BULLARD, E. C., A. E. MAXWELL, and R. REVELLE. *Advances in Geophys.*, **3,** 153 (1956).
11. VON HERZEN, R. *Nature*, **183,** 882 (1959).
12. BIRCH, F. *Bull. Geol. Soc. Amer.* **60,** 483 (1958).
13. SLICHTER, L. B. *Bull. Geol. Soc. Amer.*, **52,** 561 (1941).
14. BIRCH, F. *J. Geophys. Research*, **57,** 227 (1952).
15. UFFEN, R. J. *Trans. Am. Geophys. Union*, **33,** 893 (1952).

16. Simon, F. E., *Nature*, **172,** 746 (1953).
17. Gilvarry, J. *J. Atmosph. Terrest. Phys.*, **10,** 84 (1957).
18. Strong, H. M. *J. Geophys. Research*, **64,** 653 (1959).
19. Verhoogen, J. *Trans. Am. Geophys. Union*, **32,** 40 (1951).
20. Verhoogen, J. *Trans. Am. Geophys. Union*, **36,** 866 (1955).
21. Munk, W., and R. Revelle. *Monthly Notices Roy. Astron. Soc.*, **6,** 331 (1952).
22. Jacobs, J. J. The earth's interior, in *Handbuch d. Physik*, vol. 47, pp. 364–406. Berlin, Springer (1956).
23. Bullard, E., and H. Gellman. *Trans. Roy. Soc. (London)*, A, **247,** 213 (1954).
24. Hide, R. The hydrodynamics of the earth's core, in *Physics and Chemistry of the Earth*, vol. **1,** pp. 84–137. London, Pergamon Press (1956).
25. Verhoogen, J. *Geophys. J.*, **4,** 276 (1961).
26. Lubimova, H. A. *Geophys. J.*, **1,** 115 (1958).
27. Gutenberg, B. *Trans. Am. Geophys. Union*, **39,** 486 (1958).
28. Gutenberg, B. *Bull. Seism. Soc. Amer.*, **48,** 301 (1958).

REFLECTIONS ON THE CULTIVATION OF SCIENCE*

By Paul Delahay
Louisiana State University

The cultivation of science is a highly personal experience strongly dependent on human factors. Scientists, in general, are hesitant to discuss this experience and yet, in their rare moments of introspection and aside from any deeper philosophical preoccupations, they certainly must ponder their motives for engaging in scientific work, their methods of inquiry, and the evaluation of their achievements. Let us examine these three aspects of the experience of science.

Motives for Scientific Inquiry

Three groups of motives for scientific inquiry are readily distinguished: intellectual curiosity and aesthetic motives; utilitarian ends and improvement of the human condition; and personal factors such as ambition and escape from daily life. These motives are combined with varying degrees in an

---

* Some of this material was combined in the lecture with examples from my research work. Technical material is deleted from this text to simplify matters and preserve the continuity of thought.

individual, and the predominance of one motive varies with the age of the individual. I shall consider the first two groups of motives.

Intellectual curiosity versus utilitarian ends. Preponderance of one of the two motives, intellectual curiosity and interest in utilitarian ends, has varied with periods and races throughout history. Science began with rudiments of technology among primitive peoples [1]. Babylonians and Egyptians, in a way, practiced science for its useful ends; for example, rudiments of geometry were essential in the surveying of land after the annual floods by the Nile [2]. We must, however, turn to the Greeks for the cultivation of science for its own sake and the full development of abstract thinking. In contrast, the practically minded Romans were more interested in engineering than in speculation and, if one excepts the period of Arab glory (seventh to twelfth century), the cultivation of science largely vanished until the Renaissance and the subsequent Scientific Revolution. This is not to say that rationalism had died during the Middle Ages. On the contrary, extreme rationalism put its metaphysical stamp on every discussion and hampered the progress of science. The Scientific Revolution was an age in which scientific curiosity was not channeled into practical applications, but this situation did not prevail for long with the advent of the Industrial Revolution because of the increasing interplay of science and technology. Scientists also lost their status as amateurs, and the professional scientist made his first appearance.

At present, the usefulness of science is emphasized more than ever, and intellectual curiosity is not always recognized as a prime motive for scientific inquiry. One often hears about the importance of "basic" research "because useful results always follow," but is this detached intellectual curiosity? Of course, science is studied intensively for its own sake at the present, but it is not impossible that incessant justification of science on pragmatic grounds may weaken intellectual curiosity.

Applied science no longer needs justification to a public accustomed to its benefits, and the label of applied science sometimes serves as a disguise for research which can make no claim to be of applied nature. It is not impossible, as Polanyi [3] suggests, that "people may perhaps continue indefinitely to cultivate pure science, while professing a theory of science which exposes this occupation as a pretence or condemns it as an abuse." At worst, we may even question whether the cultivation of science will not disappear some day [3]. After all, it largely disappeared with those able administrators and engineers, the Romans.

On the nature of scientific curiosity. It is convenient to refer to scientific curiosity as a major motive in the cultivation of science. But, what is scientific curiosity? To answer this question let us consider the general goal of science divorced from apparent utilitarian ends.

Science is an attempt to put some order in our views about nature. By means of correlation and unifying ideas, science decreases the number of questions to which we seek answers. Let us examine as an example the study of rates of chemical reactions, i.e., chemical kinetics. Reaction rates vary over a very wide range, and mere compilation of rates would be a formidable and certainly not enlightening task. Great progress is achieved, however, when it is noted that reaction rates obey simple mathematical formulas in terms of the concentrations of reacting substances, For instance, the rate of reaction between two substances often is proportional to the product of the concentrations of reactants raised to some power, $\frac{1}{2}$, 1, 2, and so on. Thus, the introduction of empirical rate constants enormously simplifies matters. Chemical kinetics becomes organized, and the number of questions about reaction rates to which we seek answers is considerably reduced. Our next step is now to correlate qualitatively and quantitatively, if possible, rate constants with the nature of reacting substances. This brings about the correlation between rate constants and chemical structure. More illustrious examples than that of chemical kinetics can be quoted: Newton's mechanics correlates numerous observations about celestial bodies; the Schrödinger equation correlates energy levels in atoms and molecules; Darwin's ideas unify a vast body of observations.

We can now give an answer to the question raised: What is scientific curiosity? It is an urge to increase and simplify our knowledge about nature. This deep-rooted urge, perhaps, was ultimately a defense against the fear of the primitive man in his fight for survival. Ortega's forceful words may well be recalled here [4]:

"... man is not born in order to dedicate his life to intellectual pursuits, but vice-versa; immersed willy-nilly in the task of living, we have to exercise our intellects, to think, to have ideas about what surrounds us; but we must have them in truth, that is, have our own. Thus life is not to be lived for the sake of intelligence, science, culture, but the reverse; intelligence, science, culture have no other reality than that which accrues to them as tools for life. To believe the former is to fall into the intellectualistic folly which, several times in history, has brought about the downfall of intelligence because it leaves intelligence without justification at the very moment of deifying it and asserting that it is the only thing which does not need justification. Thus intelligence is left in the air, rootless, at the mercy of two hostile forces: on the one hand, the bigotry of culture; on the other, insolence against culture."

Applied science. Applied science is cultivated for altruistic or lucrative reasons, and it needs no other justification than the improvement of the human conditions and financial gain. Since applied science depends on basic science, the question is often raised whether industry should respond to the call for basic research. (The same question applies to medicine and agriculture.) There is no question that industry should support basic science—and it does to a limited extent—but should industry contribute itself to basic research? There is a current vogue in industry for basic research, but in many instances this is a concession to the fashion of the day to designate fairly long-range applied research. Although there is a close interplay of science and technology, industry has not the primary function to develop science and it needs no apology for not engaging in basic research if this type of research is not to its interest. Whether industry always understands its own long-range interests is of course another question. Actually, only relatively few industrial laboratories have truly outstanding scientific achievements to their credit. Fisher's recent survey [5] of the distribution of scientific publications among industrial research laboratories is revealing in this respect.

The Ways of Scientific Research

The two-way path from observation and experiment to generalization. Scientific interpretation begins either with observation or experiment (i.e. controlled observation) or both, or it begins with generalization. In the former case, description and interpretation by abstraction or by an empirical approach follow observation (with or without experiment) and ultimately lead to generalization. In the opposite process, a law which is known or postulated leads to experiment by deduction. Abstraction can of course be bypassed by empirical interpretation, but satisfaction with this approach, although justified for problems defying abstraction in the present state of knowledge, is a sign of intellectual laziness or deficiency. Statements such as "no time for theory" or "let us get our hands dirty" often reveal ignorance coupled with a desire for rapid results in individuals who "get things done."

When I referred above to the two-way path from observation (or experiment) to generalization I did not imply that a great mass of data is necessarily needed. This may be the case in the purely descriptive parts of certain sciences such as botany or zoology, but many of the great ideas of science did not emerge as a result of patient compilation of new data. These ideas were conceived to account for existing experimental knowledge [6]. Thus, Fermi postulated the neutrino to account for the emission spectrum of β-particles some twenty years before experimental evidence was presented for the existence of this particle [7].

The key stage of abstraction in analysis or deduction is a re-creation process of simplification. Intuition plays here the essential role of guide. The process of abstraction is not unlike creation in the visual arts. The painter or the sculptor must also re-create the object of this study, and works that merely attempt a servile representation do not withstand the test of time. The manner of re-creation depends on the prevailing style, which in its turn reflects the spirit of the time. Scientists also have their styles and fashions!

Re-creation in science is based on models that are mechanistic, mathematical, or a combination of both. The use of models has been very well summarized by Butler [8]. "The actual phenomenon is thus replaced by a simplified model which behaves according to laws and principles which are clearly formulated and of which the consequences can be worked out. If the predictions deduced from the model, behaving in accordance with the laws, are successful, the model is regarded as a correct representation of the phenomenon." The purely mathematical steps, if any, in the deduction process are self-consistent (barring, of course, errors), and the validity of conclusions rests entirely on the premises. This point is overlooked when the complete reasoning is endowed with the rigor which the purely deductive steps may solely possess. Mathematics is "the most original creation of the human spirit" [9] but translation of nature into its language is indeed full of pitfalls.

The extent of visual representation in models greatly varies with the period of science, the nature of problems, and individuals. Mathematics has its geometers and analysts; many chemists have a bias toward either thermodynamics or kinetics. The classical physics of the nineteenth century rested heavily on mechanistic interpretations (excluding Maxwell's work, for instance) while modern physics is devoid of them and is divorced of intuition based on common day-to-day experience.

Science, however, regardless of the details of its methods, has one feature which emerges among all others, namely the deep-rooted conviction of scientists that natural phenomena can be interpreted by mathematical constructions. This a priori esthetic conviction, although it led on occasions to unbridled rationalism with the Greeks, became particularly fruitful when it was tempered by the empiricism of the Scientific Revolution. This belief pervades science from Pythagoras and his interpretation of nature based on numbers, to Kepler and his formulas for the motion of planets, and to Schrödinger and his equation for energy levels in atoms and molecules.

Selection of ideas. How does the scientist develop ideas and how does he select those worth pursuing? Poincaré's famous account of his discovery of Fuchsian functions is an answer to the first part of my question. In Poincaré's words [10]: "... the idea came to me ... that the transformations I had used to define the Fuchsian functions were identical with those of non-Euclidian geometry. I did not verify the idea; ... but I felt a perfect certainty."

Two essential features of the scientific method appear in the account of Poincaré. First, a "search for unity in hidden likeness" [11]; and second the observation that "Scientists—that is, creative scientists—spend their lives in trying to guess right" [12]. Supporting examples for the first point are immediate: the creation of analytical geometry by Descartes by the fusion of geometry and algebra, the correlation between optics and electromagnetic theory by Maxwell, the correlation of apparently unrelated areas in mathematics by the theory of groups.

Familiarity with previous knowledge, quite obviously, is essential in the process of correlation, and many opportunities have been missed by lack of this knowledge. Consider, for example, the development of gas chromatography. Martin and Synge [13] in their initial paper on partition chromatography (1941) suggested extension of their ideas to the separation of substances in the gaseous state. This suggestion was not taken up by the numerous readers of this fundamental paper, perhaps because most readers saw direct applications in their field of interest (often biochemistry) and were not concerned with the more remote possibility of gas separation. At any rate, Martin[14] finally (1952) followed his own suggestion, some ten years after having made it, and developed gas chromatography in collaboration with James. The countless applications already in the literature testify to the usefulness of the method.

The possibility of guessing right—the second characteristic of the scientific method as I indicated above—is, at the highest level of science, the apanage of genius. A typical example is offered by Galois, a French mathematician (1811–1832) who, in his brief life, advanced ideas which by their power rank him among a few of the greatest mathematicians. Hadamard [15] recounts (for further details see Sarton [16]) that the night before the duel in which Galois received a mortal wound, he wrote his mathematical testament and advanced ideas far beyond his time. How Galois guessed right on ideas which had not been developed at his time can only be ascribed to the intuition of genius. Now this is easy to say but I have simply modified the question by answering it with the magic word of genius. I must leave the answer to others (see Polanyi [3, 17] for instance).

Ideally, the scientist directs his thoughts toward problems he deems significant and regards within his competence. His choice in actuality is also dictated by other factors such as the interest of his time and chance. The interests of a scientist's time have a profound influence as Bronowski [18] points out for the three following examples: *1*) Newton turned to astronomy in a society preoccupied with navigation; *2*) Faraday devoted much of his life to the relation between electricity and magnetism in a society searching for new sources of power; *3*) Professor Wiener developed cybernetics in a society with increasing needs for automation. Let me add another more modest example: the current interest in fast reactions in chemical kinetics is natural in an age concerned with rockets and jet engines. It is not the practical ends resulting from these studies that incited Newton, Faraday, or Wiener, but the general interest of their time oriented them in their selection of fields of inquiry. Likewise, the chemist interested in fast reactions does not limit himself to processes directly relevant to practical applications but also considers fast reactions in solution—fast electrode processes, for example.

The climate of the time may also have an adverse effect on the scientist, that is on the pioneering scientist. His work may be overlooked for many years because of the lack of interest by his contemporaries or the lack of adequate theory or equipment. Chromatography is a classical example in this respect [19]. This method for the separation of minute quantities of substances from mixtures was originally developed by the Russian botanist Tswett at the beginning of the century. Tswett's work remained buried in the literature for almost thirty years because chemistry at that time was oriented in other directions. It was only when the need for separation methods arose, mostly in biochemistry, that chromatography received the attention it deserves. Likewise, non-Euclidian geometry did not receive immediate acclaim as one of the greatest achievements of the nineteenth century because it did not fit in the spirit for visualization of that century [20].

In addition to the climate of the time, chance often affects the scientist's selection of problems. The history of science amply proves this point (cf. Taton's book [19]). Suffice it to say that Fortune helps only those who, by their acuteness of observation and willingness to depart from accepted ideas, are ready to receive her gifts.

The originality of an idea has an important influence in determining the scientist's choice of problems. Originality exerts a fascination on man, and it is as essential to science as to art. But the idolatry of originality has its dangers: the scientist striving to be original may substitute originality for significance. How much effort has been spent in the brilliant solution of difficult, but quite unimportant problems?

Loss of universality. The expansion of scientific knowledge obviously has resulted in increasing specialization. We have lost the universality of the Greeks [21]. What is more serious, we have become accustomed to shun meditative thinking outside our field of specialization (cf. Barzun [22] for instance), and we have developed "minds in a groove" [23]. By the loss of universality man has dulled and distorted his sense of values. Scientists must limit themselves to relatively small areas of science because of the immense scientific output, but many of them inflict upon themselves, as gifted individuals, an unconscious punishment by a lack of genuine appreciation for things outside the realm of science or even a very small part of science. (What of *les littéraires* with an appalling ignorance of the essential contribution of science to the history of ideas?) There are many reasons for such a lack of universality, the least valid of them being perhaps that professional activities leave little free time. Far from me the thought that knowledge must be gained for the pedantic and doubtful pleasure of displaying it or that encyclopedic, bookish knowledge must be acquired. The lack of eclectic choice only results in a confusion of mind that is worse than ignorance. Our perspective of life depends on the breadth and depth of our interests, and every serious effort to broaden our vistas should be applauded. But, let us beware of those who speak too much of culture!

Evaluation of Scientific Achievement

Criteria for evaluation. At the onset we must make a distinction between the evaluation of scientific achievement, which is quite impartial, and the evaluation of scientists, which is colored by personality and other human factors when contemporaries are involved. We evaluate scientific achievement as we judge a painting, that is on its own merit and without interference by recollections about the painter's life. Posterity makes few mistakes, but our judgment of contemporary scientists is quite biased and often is not based on the scientist's work but rather on what we *hear* about this work. This collective evaluation is, in general quite correct but it nevertheless is uncritical. Caplow and McGee [24] find the same uncritical approach in university hiring practices.

Our evaluation also is biased by our own approach in science. The specialist of organic preparative chemistry, in all honesty with himself, may have little real appreciation for theoretical chemistry though he is fully aware of its significance; and, conversely, the theorist may be disdainful of his colleague at the bench. Science is not free from internal nationalism.

Which criteria do we apply in the evaluation of scientific achievement? The answer is quite simple for the general public which identifies science with technology and medicine: the main criteria are usefulness and, to a lesser extent, the magnitude and cost of the effort. This is not to say that criteria applicable to pure science do not hold in applied science, but usefulness is the dominant factor. Conversely, scientists nowadays are tempted, not infrequently, to adopt the criteria of magnitude and cost of effort in evaluating their work. Tuve [25] had some sharp words for the empire-builder type of professor at the recent Basic Science Symposium in New York: "... professor's life nowadays is a rat-race of busyness and activity, managing contracts and projects, guiding teams of assistants, bossing crews of technicians, making numerous trips, sitting on committees for government agencies, and engaging in other distractions necessary to keep the whole frenetic business from collapse.... Too many of our academic leaders have chosen this pattern of activity and personal power in preference to the quieter and more difficult life of dealing with ideas and scholarly initiative." Indeed some of us behave as if we had forgotten that scientific achievement should be judged by the significance for the advancement of knowledge and the originality and boldness of conception.

The economy of means also provides a criterion of evaluation. Is Whitehead not carried away when he writes about the Lagrange equations for motion [26]: "The beauty and almost divine simplicity of these equations is such that these formulae are worthy to rank with those mysterious symbols which in ancient times were held directly to indicate the Supreme Reason at the base of all things." We admire the *apparent* simplicity of these equations just as we admire the apparent simplicity of a drawing by Matisse or some of the painted Greek vases.

A last word about originality and boldness of conception: our admiration for them has led to their sham emulation in the current fashion for "creative activities." Books are written in the spirit of "do it yourself" about creative thinking and other activities. These books have their points, but I am quite sure that some of them would be the delight of a sarcastically-minded reader.

Achievement in relation to knowledge of the time. The significance of scientific work is an evaluation criterion which, however, must be interpreted in relation with the time of the work. An investigation which is rightly regarded as of great importance may lose much of its significance a decade or two later. Obviously, there is progress in science but not in the quality of scientific achievement and, for that matter, of artistic creation. Who would submit that Einstein is superior to Newton? Great intellectual achievements of past science only *appear* easy in comparison with the complexity of present day science. The boldness and originality of an idea wear off as we become accustomed to it. Thus, the examiners of de Broglie's dissertation—and they were eminent (J. Perrin, Langevin, Mauguin, and Cartan)—while they admired the depth of thought of the candidate were quite sceptical about the significance of his famous equation [27]. Nowadays, the student of elementary physical chemistry takes this equation in his stride. In this respect, textbooks, in general, do not convey the human factors of discovery nor do they allude to the mistakes that have marked the development of science. Many of the greatest of the trail blazers of science—not so much mathematicians, however, because of the very nature of their field—made errors (see Polanyi [28] for actual instances). The development of science is a succession of human events and, just as in history, the grand pattern often emerges well after the events. We tend to straighten the sometime tortuous path of scientific discovery by dissociating the great scientist from his time and isolating him on a pedestal. We may overlook that the subject was ripe for discovery by his time and we may forget his predecessors. When we think of Newton we should not overlook Copernicus, Kepler, and Galileo.

Scientific honesty. One remarkable trait in scientists, scientific honesty, is generally not admired because it is taken for granted. A scientist of doubtful character may not hesitate to borrow ideas ruthlessly without acknowledgment, but he will not falsify or invent data in the reporting of his work. He would otherwise face almost inevitable punishment. The rules of the game cannot be flouted in science, and scientific reports have a stamp of authenticity unequaled in many other human documents (cf. Bronowski [11, 29]).

Without honesty in the cultivation of science there cannot be the reward of elation accompanying successful work. The dishonest scientist robs himself of one of the principal sources of happiness in man, as Pauling [30] calls the joy of discovery.

REFERENCES

1. SARTON, G. *Introduction to the History of Science*, vol. I, pp. 3–6. Cambridge, Harvard Univ. Press (1952).
2. HULL, L. W. H. *History and Philosophy of Science*, pp. 5–6. New York, Longmans, Green (1959).
3. POLANYI, M. *Personal Knowledge*, p. 181. Chicago, Univ. of Chicago Press (1958).
4. ORTEGA Y. GASSET, J. *Man and Crisis*, M. Adams, Trans., p. 112. New York, Norton (1958).
5. FISHER, J. C. *Science*, **129,** 1653 (1959).
6. POLANYI, M. *Science, Faith and Society*, p. 14. London, Oxford Univ. Press (1946).
7. HANSON, N. R. *Pattern of Discovery*, pp. 125, 218. London, Cambridge Univ. Press (1958).
8. BUTLER, J. A. V. *Science and Human Life*, p. 106. New York, Basic Books (1957).
9. WHITEHEAD, A. N. *Science and the Modern World*, p. 29. New York, Macmillan (1935).
10. POINCARÉ, H. *The Foundations of Science*, G. B. Halsted, Trans., p. 388. New York, Science Press (1929).
11. BRONOWSKI, J. *Science and Human Values*, p. 23. New York, Julian Messner (1956).
12. Ref. 3, p. 143.
13. MARTIN, A. J. P., and R. L. M. SYNGE. *Biochem. J. (London)*, **35,** 1358 (1941).
14. JAMES, A. T., and A. J. P. MARTIN. *Analyst*, **77,** 915 (1952).
15. HADAMARD, J. *The Psychology of Invention in the Mathematical Field*, pp. 118–121. New York, Dover (1945).
16. SARTON, G. *The Life of Science*, pp. 83–100. New York, Henry Schuman (1948).
17. POLANYI, M. *The Study of Man*, p. 19. Chicago, Univ. of Chicago Press (1959).
18. Ref. 11, pp. 15–16.
19. TATON, R. *Reason and Chance in Scientific Discovery*, A. J. Pomerans, Trans., pp. 159–162. New York, Philosophical Library (1957).

20. Ref. 2, p. 212.
21. SCHRÖDINGER, E. *Nature and the Greeks*, p. 12. London, Cambridge Univ. Press (1954).
22. BARZUN, J. *The House of Intellect*. New York, Harper (1959).
23. Ref. 9, p. 282.
24. CAPLOW, T., and R. J. MCGEE. *The Academic Marketplace*. New York, Basic Books (1958).
25. TUVE, M. A. *Saturday Review*, p. 49, June 6 (1959).
26. Ref. 9, p. 91.
27. MAUGUIN, C. In *Louis de Broglie, Physicien et Penseur*, pp. 431–436. Paris, Albin Michel (1953).
28. Ref. 6, pp. 75–80.
29. BRONOWSKI, J. In *New Knowledge in Human Values*, A. H. Maslow, Ed., pp. 52–64. New York, Harper (1959).
30. PAULING, L. Foreword to *Moments of Discovery*, G. Schwartz and P. W. Bishop, Eds., vol. 1, p. vii. New York, Basic Books (1958).

THE PROBLEM OF THE SCIENCES AND THE HUMANITIES
A Diagnosis and a Prescription*

By Harold G. Cassidy
Yale University

A major cause of the developing cultural fission which we see as the problem of the sciences and the humanities is lack of effective communication between scientists and humanists. It is toward ways of improving communication, and bringing scientists and humanists more effectively together, that this essay is addressed. We present not only a diagnosis, but a prescription.

It may be asked why a subject of this kind should be brought before The Society of The Sigma Xi. One reason is that as companions in zealous research we must be concerned with the fact that the products of our research are controlled and used far more by nonscientists than by scientists. An examination of the roster of the Congress of the United States, of officials of industrial concerns, and of advertising managers shows this. Furthermore, it is through the arts that science

* Based in part upon a book, *The Sciences and The Arts: A New Alliance*, New York, Harper and Brothers, 1962.

must be interpreted to the layman who, to some extent, controls the products of our research through his votes and his dollars. He will not read the *Journal of the American Chemical Society* but he will read and be affected by the host of newspapers, magazines, and books which compete for his leisure time. It is a responsibility of the educated scientist to help in this interpretation, as it is the responsibility of the humanist to make the effort necessary for it, because otherwise an anti-intellectualism, which begins with misunderstanding and fear, can swallow up both science and art.

There is another extremely important reason for dealing in this place with this subject. It has to do with "speaking up." We are an educated part of our community, and it is our responsibility to have opinions on matters which affect us, our local community, and our nation. Unfortunately, our education, by and large, has not been geared to serve us in this way. The factual basis in science and history, the wisdom of philosophy and religion, have been skimped in favor of incredibly time-wasting and intellect-degrading puff-courses in "driver education," "social adjustment," and the like. The degradation of our language that goes on apace via the sophistries of the advertiser, who dares not speak precisely and clearly because then his product is seen to be essentially like his competitors': that it comes in pint bottles instead of the "big half-quart size," has left us subdued in the presence of questions, the answers to which require the most careful analysis, and the most precise communication. And the cult of the expert, coupled with our own specializations which do, indeed, show us how difficult the problems of this world can be, leaves us unwilling to speak "out of our field."

This situation is dangerous for two reasons. One is that it is divisive: we are silent when we should speak—and silence can separate people as effectively as outright conflict. But our silence does not dispose of the problems. It merely creates a vacuum into which rush all manner of less inhibited "experts," often of doubtful repute, often with special axes to grind, often of extraordinary narrowness of knowledge and outlook. This is the other reason.

We venture to offer a prescription that will help us to correct this situation. It will not be an easy prescription to follow because it involves the attainment in each of us of a reasonably balanced position with respect to the sciences and the humanities. Extreme positions are relatively easy to take; one merely fails to see all contrary evidences. Extreme positions are relatively easy to maintain; adversaries can come from only one direction. But the rational, balanced view is hard to take, because values must be weighed judicially. It is difficult to maintain, because it will be attacked from all sides. Nevertheless, it has a certain nobility, rationality, and viability which must recommend it to us; and it gives us hope for the future.

Lest it be thought that we are here constructing bogeys, let me quote from two acquaintances. One was a minor university official who told me, "I've never had a course in science, and I'm proud of it." Before the humanists blush too deeply, and the scientists be too shocked, I quote the other—a professor of physics. Speaking of the college curriculum, he said, "The sooner we get rid of all this humanistic tripe, the better." (He was referring to literature and history.)

In order to offer an effective prescription we must make a diagnosis. This we will do by first offering two concrete examples which show how a poet and a physicist have communicated ideas of great importance. After discussing these briefly, we proceed to abstract from them. This leads us into a discussion of the structure of the university, and out of this arises our prescription in a natural way.

To deal with such complicated matters in a little space requires a directness of approach that will prevent us from citing minute detail and variety of opinion. Also, we have taken our concrete examples from a poet and a physicist since, in a very real way, they represent the clearest contrasts between humanist and scientist—much clearer, for example, than those between a historian and a psychologist. What we have to say, however, will be applicable to any humanist or scientist.

Diagnosis

As concrete examples we have chosen a few lines from a longish poem "Einstein," written in 1929 by Archibald MacLeish* and a formula from a work written 250-odd years ago by Sir Isaac Newton. (Both are taken out of context.)

"... He lies upon his bed
Exerting on Arcturus and the Moon
Forces proportional inversely to
The squares of their remoteness,
and conceives
The universe. ..."

$$F \propto \frac{m_1 m_2}{s^2}$$

Now the first point that strikes one is the difference in appearance of the two. The poetry, written in English, can easily be read; the formula is obscure to anyone who has not studied physics. The formula says that there is a force (F) which is proportional to ($\propto$) the products of the masses of two objects (m_1, m_2) and inversely proportional to the square of the distance (s^2) between them. This formula may be rewritten for purposes of calculation as $F = Gm_1m_2/s^2$, where G is a constant of proportionality which adjusts the units used. There are certain limitations on this equation: it is approximated only if the distance s is *large* compared with the diameters of the objects, unless both objects are spherical. However, the application to the poetry is clear. Since m_1 and m_2 are not specified, *one* could refer to Einstein and the *other* to Arcturus and, since, compared with their diameters

* Houghton Mifflin Co.

the distance between them is large (it is about 32.6 light years), the formula would hold to a good approximation. Thus what the poetry says agrees, as a special example, with the formula.

Now, a case could be made that no two people get exactly the same message from the poetry. This is because each important word is festooned with connotations, many of them quite different for different people. Thus the star Arcturus must surely carry different connotations for people from the Lone Star State, from those it carries for people from the other forty-nine. Some people are old enough to attach to Arcturus the recollection that a beam of light from this star was used to open The Century of Progress World Fair in Chicago, in May 1933. The connotations adhering to "moon" are known to change with the age and sex of the person.

On the other hand, it is very likely that all physical scientists understand essentially the same message from the formula as long as it is being used as a scientific communication. This is because the denotative symbols are essentially naked of connotations.

The formula may be used in another way, of course. The scientist may see it as a manifestation of the power of the human mind and imagination. He may ponder on the implications of Newton's creative insight. In this case he is using it not scientifically, but poetically. In this use it may *function* as a poem, or other work of art, though it does not have that *appearance*. In this use, different scientists will read different messages into the equation.

A test we might apply to our two texts is that of translation from one language to another. If the previous statements carry conviction, then it must readily be agreed that the translated poem is unlikely to convey the translator's or the poet's thoughts exactly to his reader. The formula scarcely needs translation, since the symbols are part of an international scientific language, and the formula itself defines the force as gravitational. The contrast brought out here partly explains the international character of science and the fact that nationalism feeds on the poetry, and particularly the songs, of a country.

To the poet the formula may seem cold, while his poem pulses with warmth and life. The scientist does not deny this. He is wise if he points out the price each pays for what he gets. By paying the price of lack of warmth, the scientist achieves wide communicability of a very precise and quantitative message. The poet achieves his warmth, but communicates a quite private message to whoever can hear. It is largely a matter of emphasis, of course, since the way the world is constituted allows us to attain no absolutes except in trivial cases. The poem has its denotation, but this is the least of its messages; the formula has its connotations, but these are essentially excluded in its scientific use.

The question may be raised: what kind of truth does each communicate? A good approach to an answer has been given by John Hospers. He characterizes the truths of science as "truth-about" the objective world. This is a kind of truth which is publicly verifiable. Anyone with the requisite training can check Newton's formula to a reasonable approximation by carrying out one of the famous experiments or calculations that support it. Or, he can devise a different, hitherto untried, experiment and, provided he meets the limitations of the formula, he will confirm it. The formula says, in effect, that this is the way certain aspects of the world (masses of objects, distances between them) are related to each other.

The arts communicate truth of an essentially more private nature, "truth-to." This kind of truth is validated by the communicant out of his own experience. The composer, artist, or author produces a work which, as John Dewey has said, creates participation of the listener, viewer, or reader in the artist's experience. Perhaps the artist has created a character who is recognized by his readers as "truer to life than life itself." Thus out of his own experience, or some felt need which is satisfied by the work, comes a personal validation of the work as true.

These differences between art and science are closely linked to the tools for communication used by artist and scientist. In science the drive always is in the direction of quantitative statements, the goal being relationships that can be stated by means of equations or formulas of the type of our example. This equation states *a ratio*. As such it reaches the acme of precision. The constant G is present to rationalize the units of s and m. Given G and the values of any three of the four quantities, the fourth can be calculated with complete *mathematical* certainty. Any uncertainty resides in the values of the quantities themselves. The ratio (or formula) is thus a very powerful mathematical tool.

The majority of relationships which are studied by scientists cannot be stated in terms of ratios of the type illustrated. They can be put into mathematical form, but the equations contain adjustable parameters. Now the other important tool of the scientist is *analogy*. An analogy is a relationship of likeness, as is the ratio, but it is an imperfect ratio. In a sense, it may be likened to an equation with many adjustable parameters. Such an equation may allow a number of different theoretical interpretations, and one problem of research is to find out which, if any, of these is best—the criteria of "best" being correspondence to experiment, and fruitfulness or correct predictions (among other criteria). Analogy is the principal tool for communication of relationships (as distinct from mere description of phenomena) in the biological, social, and related sciences. But the drive is always toward quantitative statement.

In the arts, too, one of the principal tools of communication is *analogy*. Yet there are many things which cannot be communicated even by this tool. These are the things that verge upon the ineffable. They must be communicated and, when in our need we feel that we must, we turn to the greatest artists who have from time to time succeeded in this nearly impossible task. They have invoked the aid of *metaphor*. We quote them, or point to their work, and thus substitute their genius for our fumbling. Metaphor is the most powerful tool of the humanist. "Its efficacy verges on magic," says Ortega, "and it seems a tool for creation which God forgot inside one of His creatures when He made him." By the aid of metaphor it is possible to attain a qualitative precision which is *functionally* analogous to the quantitative precision achieved by the use of ratio.

We see, then, that the emphasis in science is on ratio and analogy, whereas that in the humanities is on metaphor and analogy. But the humanist does not hesitate to approach the use of ratios—as in stylometric analysis—if it suits his purposes; and the scientist nowadays is sufficiently secure that he can turn to metaphor, as when the nuclear physicist speaks of "magic numbers," not knowing, at first, what they signify.

Definitions

A university catalogue lists, already defined in it, the various disciplines that are taught. If there is doubt about any one of them, one can go and ask the people who are members of the particular department, what they do. Or, one can observe what they teach, and what they publish. Because each department has a budget, its definition as a department can be further certified. We group departments: physical sciences include physics, chemistry, geology, astronomy; life sciences include botany, microbiology, zoology; on the border between physical sciences and life sciences would fall biochemistry, biophysics, ecology; and so on. This grouping is done for convenience of representation, and because straight enumeration will not accomplish our purpose, since the individual disciplines cannot be arranged in an entirely logical sequence; they did not develop in that way. So, by grouping them we can avoid trying to make minute distinctions while preserving the larger differences.

We then proceed to define the sciences and the arts (or, humanities) in the following way. We arrange these groups in a logical sequence around the perimeter of a circle, as is

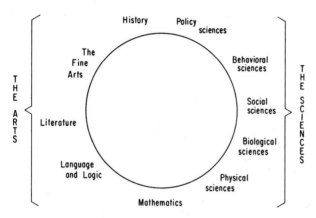

FIG. 10. Definition of the arts (the humanities) and the sciences.

shown in Figure 10. The two arcs opposite each other comprise the arts and sciences, as labeled in the figure. It can be seen that a certain internal logic is displayed by the figure. On the science side, there is (by and large) a sequence from inorganic to organic; from preoccupation with statistical aggregates and chance events to concern with phenomena displayed by individuals, and small groups of individuals, and events controlled by individuals and small groups. On the arts side there is a sequence from discursive to presentational means of communication, with an emphasis at the upper part of the arc on the more local or nationalistic expressions, which verge then into history.

We explicitly do *not* wish this figure to be taken too literally. Disciplines across from each other are not opposite, or opposed in any real way, nor is the width of the sectors given to a discipline meant to imply anything but typographical convenience. What we do explicitly imply is that the whole represents a connected *field*, without natural barriers; that in defining the arts and sciences in this way we do convey a kind of oppositeness between them in the large, in that there are areas of knowledge and experience which in actual fact each takes as its own; and that there are disciplines which belong to both areas: mathematics, one of the oldest of the humanities and newest of the sciences, and history, which finds itself sometimes under the Dean of the Humanities, sometimes under the Dean of the Social Sciences.

This preliminary diagram, Figure 10, does not include a number of other disciplines which are taught in the university, such as engineering, philosophy, law, medicine. In order to complete the picture, which will be that of the knowledge and experience taught in the university as a whole, we must define and discuss three more terms.

In *all* disciplines of the university which have intellectual content there are practiced three kinds of activities: analysis, synthesis, and reduction to practice. *Analysis* is the activity of gathering data, describing things as they are, collecting and recording instances, making lists, and so on. Now an intelligent person cannot go very far in this kind of activity before he begins to see patterns in his data, likenesses in his descriptions, similarities among the separate instances. As this occurs, the essentially analytical activity goes over into a synthetic one. *Synthesis* is the activity of finding connections between the data made available by the analytical activity, making hypotheses and theories, and developing laws. In short, synthetic activity is a generalizing activity. It involves abstraction, for the general statement is at a higher level of abstraction than the many single instances on which it is based. But the hypothesis, theory, or law must be reduced to practice. *Reduction to practice* is the activity through which a return is made from the general principle to the single instance. It may be carried out to see if a predicted fit is obtained. It is then the experiment designed to test a theory. It may be the examination of a document to see whether it meets some critical test. It is also the creative activity by which the work of art or science is produced.

There are two features of these activities which we must emphasize. One is that it is difficult in many instances to classify some activity clearly as one or the other kind. We have had to separate them for the purposes of definition. But when we try to apply our concepts to actual cases we find that we are dealing with a mixed situation, and we have to recognize not pure analysis or pure synthesis, but *emphasis* more in one direction or the other.

The second feature is that these activities may be carried out at many levels of abstraction. Thus theories may be the subject of analysis, which might then lead to hypotheses or theories about them, and these higher-level generalizations may be tested by reduction to practice. Or, methods of reduction to practice may be collected and examined with a view to the heuristic enterprise of developing the method of the methods; that is, theory at a more abstract level.

The Structure of the University

The foregoing discussion and definitions permit us to illustrate, in a manner that must not be taken as more than an analogy, a concept of the structure of the university. This is shown in Figure 11. The figure is idealized as a sphere, not because the ancient Greeks considered this to be the most perfect figure in solid geometry, but because it helps to avoid invidious comparisons. The circle of Figure 10, used to define the arts and sciences, now comprises the equatorial belt of our sphere. Toward one pole we gather the technologies; toward the other the philosophies.

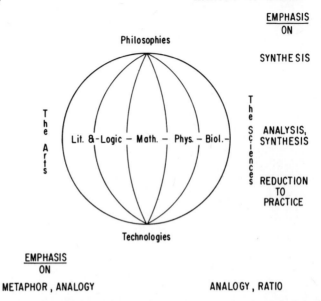

FIG. 11. Conceptual analysis of the structure of the university. The equatorial band of disciplines, shown in part, is taken from Figure 10 and extends around the sphere of Figure 11.

The technologies we define as disciplines wherein reduction to practice is emphasized. But we must insist that, like any other university discipline with intellectual content, the technologies comprise analysis and synthesis and, as a part of the university, they comprise the heuristic activity we described above: the analysis and generalization, at higher levels, of the technology. If pure reduction to practice were involved, we would not have a technology but a craft. In this definition we do not derogate the craft, but distinguish it clearly from the technology.

Every discipline in the arts and sciences has its conjugate technology. In the sciences these are often called engineering; in the arts they have special names. And it is a characteristic of the technology that it often calls upon many disciplines, which it then fruitfully combines. Thus medicine is the technology of a whole constellation of disciplines, ranging from physics and chemistry to literature and drama. Most kinds of engineering include mathematics, physics, and a number of other sciences. Law is the technology of jurisprudence, and calls upon anthropology, psychology, and much more. The technological disciplines of literature and the fine arts—the "commercial" disciplines—as also drama, call upon many kinds of knowledge and experience. We do not need to enumerate all the technologies. They can be found in university catalogues and are recognizable via our definition.

The philosophies we define analogously to the technologies. In them synthesis is emphasized. Here, too, many disciplines are likely to be drawn together in fruitful union. Here, too, we emphasize plurality. This is because the idiom appropriate to a philosophy of science cannot be transferred to a philosophy of art with any real meaningfulness—at least at present. We can understand this difficulty on the basis of our examination of the tools of communication: metaphor has its uses, and they are not the uses of ratio. Moreover, we do not find notable agreement among philosophers, and so we are led to the conclusion that each discipline, or group of disciplines, should have its philosophers, who in general should first, or concomitantly, be experts in that area.

We do not wish to imply by this diagram that at its south pole would be found some epitome of technology, and at its north the all-comprising philosophy. The reader is at liberty to speculate on these matters, of course.

What we do wish explicitly to emphasize is that the sphere of knowledge and experience comprised by the university is a whole, in principle, without barriers. We suggest also that, reading downward, synthetic activities are emphasized near the "top" part, analytic and synthetic near the equator, and reduction to practice near the lower part. *All three are found throughout* but the *emphases* differ in the different classes of disciplines. In the arts hemisphere, to repeat, analogy and metaphor, whether discursive or presentational in their utterance, are the chief tools of communication; in the sciences, analogy and ratio.

This figure gives us an idealized picture of the intellectual structure of the university. A college would be pictured in much the same way, except that fewer disciplines would be represented, and many of the technologies and philosophies would be absent. A technological institute might comprise mostly the lower half of the diagram, or in other cases might be indistinguishable from a university. One must in a given case look at what the institution is, and not what it is named, for there has lately been a tendency to change the names of many institutions from "college" or "institute" to "university" with little more than the promise of an increased budget to go with the legal designation. Such changes in name should properly be made only in response to an accomplished fact.

We have, of course, left out of consideration a great many aspects of the arts and sciences. But this essay is not meant to be exhaustive. Its chief function is to analyze the particular situation as we see it, and present this as a starting point for a suggested prescription. A great deal has been written on the subject of the cultural cleavage between humanists and scientists. Most of it, however, describes the situation and deplores it, without suggesting remedies. We grant that remedies for such a complicated group of illnesses are hard to come by, but we will proceed nevertheless to offer a prescription based on the foregoing analysis, since it is part of our argument that we must speak up, and attack this dangerous situation.

Premises

Underlying every analysis and prescription there must be premises, whether implicit or explicit. It must be assumed, for example, that the problem contains elements amenable to analysis and that an approach to a solution of it is desirable. Some of the premises such as these underlying our analysis have been obvious. Some must be stated explicitly in preparation for our final section.

We assume that scientists and humanists are concerned with the same world—even though they may look at different aspects of it and see it with different eyes. This is the basis for our image of the university, with its emphasis on wholeness and ideal lack of barriers between disciplines. We assume, then, that it is the same "whole" with which we are all concerned.

We assume that both art and science are universal in scope in that there is nothing whatsoever that may not be a subject for artist or scientist. We observe, however, that they differ in emphasis, because there are some areas of knowledge and experience which are more amenable to tools of one kind than of another. However, we must insist on universality of scope because we also insist that no one can predict a limit to the potentialities of human beings. Yet there does not seem to be any necessary reason why the works of the artist should be translatable into scientific terms, or vice versa. There may be many different ways of speaking about the same things, and one way may be more fruitful than another, depending on circumstances.

We assume, in this connection, that too great uniformity is undesirable, particularly in those areas where criteria of truth and validity are difficult to establish. This is why a universal language, if it could be achieved, would have certain disadvantages. It would be useful in much ordinary intercourse (and an approach to it is at present in existence in some areas of science) but it would be tragic if it displaced other languages. There are concepts that can be stated in one language which cannot easily be stated in another. As a general principle we must insist that variety has its uses; it allows flexibility, choice, and room for the appearance of the unpredictable.

Such room and flexibility are essential because of a related assumption; namely, that no ideology or other dogma is capable of comprehending the infinite variety, invention, and new insight of which human beings are capable. This is why we spoke of plural philosophies. One may extrapolate to some conceivable ultimate System, but it must be emphatically insisted upon that such an absolute is not attainable in fact. This is related to the success of liberal democracies in preserving individual freedoms of their citizens and in improving their lot and that of the world as contrasted to the totalitarian democracies which rest on dogma or ideologies. These latter cannot afford to recognize phenomena that do not fit their dogma, and the insecurity which results, when facts show that their dogma does not fit the world, reveals itself in tragic ways. The liberal democracies, not operating within the straight-jacket of an ideology, but working in a piecemeal manner within the limits of general principles of widely recognized validity—the Golden Rule, equality of opportunity, equality before the law, and so on—are able to adjust to the demands of progress in the sciences and the humanities with, on the whole, some success. At any rate, they have the possibility for change.

One more assumption, which probably does not exhaust all we have made, follows from the previous ones. It is that there cannot be any final solution to these problems—a solution which will usher in some state of Utopia. The passage of time brings new problems, some of which do not fit into the prevailing pattern.

Prescription

In the preceding sections of this essay we have provided the means for constructing a prescription which now naturally follows. It is a prescription for the treatment of underlying causes. The important point is that this prescription, to be effective, must be applied by *individuals*, and this means intelligent people who recognize the problems and are willing to accept the responsibility to speak out and to set an example. It is not a legal prescription. We have the laws, but this is prior to written law.

The prescription contains a number of ingredients. One has to do with mutual understanding, which is a prerequisite to mutual trust. Another has to do with fortification of a point of view which supports our actions and guides them through rational criteria of value. Still another has to do with means for applying the prescription.

We have seen that all of our intellectual endeavors are in principle understandable and capable of being connected to each other. If the scientist sees the artist depending heavily upon metaphor, that tool capable of such qualitative precision in skilled hands, he surely should be able to understand. He knows from his own work, and his human reactions to it, that there are aspects of experience that cannot be stated in quantitative terms. Even *he* uses metaphor when he is faced with the need to communicate such experience. The skilled worker, recognizing skill in another area, can afford to praise it. At the same time the humanist, if he were able truly to recognize the power of a scientific statement and the quantitative accuracy with which an idea can be communicated to *anyone* who can listen, can surely understand that here is a manifestation of interaction between human beings worthy also of praise.

As a class exercise, in a course in elementary physical science which I teach, I once asked a group of Yale freshmen to interpret the few lines of Mr. MacLeish's poem that were quoted above. These men had been studying Newton, and all of them saw that here, on the face of the verse, was a statement of a special case in which Newton's law of gravitation held, as the poet implied. But several of them (about ten per cent) went farther. They said that this was only a minor point. What the poet says is that Einstein exerted two forces; one of them, the gravitational force, is small—in effect trivial; the other, truly powerful force, was the conceptual one, through which, in "conceiving the universe," he changed our entire way of thinking.

These same students could also be impressed with the tremendous scope and quantitative precision of Newton's gravitational formula. Certainly the gravitational force exerted by Einstein's body upon Arcturus is negligibly small, yet it is comprehended by the gravitational formula. It seems to me, then, that when scientists and humanists can see these powers in each other (and the limitations to these powers), the precondition exists for mutual understanding.

Scientists and humanists (artists) are connected, too, by the fact that both make use of analysis, synthesis, and reduction to practice as they go about their work. This particular connection is a vital one, because in it lie the means for deeper understanding of each by the other, and especially because in it lie the means for evaluating the work of each by the other.

As another class exercise I require my students to write an essay in which they connect some other course or interest of theirs with my course in physical science. One of the freshmen claimed that the *only* deep interest he had was drama. How could this be connected to science? After some discussion it developed that he had recently seen "J. B.," Mr. MacLeish's play on the theme of Job. I asked him to write an essay on the problem that Mr. MacLeish would have faced had be made J. B. a scientist instead of a business man. The upshot of this demand, involving—in the process—a conference in which the student was persuaded to reduce his outline from four pages or so to a size which would, when fully written, occupy four pages, was a nice essay. Job was a scientist. He was being tested by The Lord, and as he tried to repeat the experiments of Galileo, the ball, rolling down an inclined plane gave different relations between distance and time in different experiments—or occasionally even rolled up the plane. The classical relations failed in various ways. There was an unpredictability which drove Job frantic, as he carried on his experiments. Yet the essay concluded on a triumphant note. Job would not give up. He was a true scientist, and believed in an inherent order in nature which, though obscured by the trials he was undergoing, yet could be found if he did not give up his faith. It seemed to me that this man must already understand at a fundamental level how a scientist operates in order to write such an essay.

To be able to see that both humanists and scientists rely upon analysis, synthesis, and reduction to practice, as functions, though the results are very unlike in appearance, gives further ground for understanding. But the soundest basis lies in the realm of valuation and a point of view.

The greatest value inheres in that work which brings together constructively, in the sense of conjoined analysis, synthesis, and reduction to practice, the largest area of truth and meaning. This applies to the public "truth-about" of the scientist as well as the private "truth-to" of the artist's communication. It applies to the more denotative meanings of the scientific statement as well as the more connotatively rich meanings of the humanist. One of our criteria of value then, is that wholeness and large compass, with depth, count for more—and much more—than do partialness and superficiality or narrowness.

Another criterion of value is part of this one. It is that rationality is superior to irrationality for, as Howard Mumford Jones has pointed out, it takes rationality to discover irrationality, not the other way around.

Why do we argue that wholeness and large compass with depth in any work make it more valuable than the lack of these? It is because when we see many interconnections in the parts of a system we feel more confidence in the validity of the system than when there are few. A theory in science that reaches from the atoms to the stars with reasonable fit to the data of such disparate realms gives us confidence that it carries truth which is not given by a theory that fits a few local instances. A theory of forms in art which can deal with art over a span of two thousand years gives us more confidence than a pronouncement, no matter how sonorous, about some local revolt in art.

And why do we praise rationality and meaningfulness over irrationality and meaninglessness? It is because the human race has found the former hard to come by. It requires effort in the same sense that staying alive requires effort. It is part of the grand pattern in which we see living beings developing in constant struggle against the degradative forces that lead to death; in which we see human beings lifting themselves, ever so slowly, and with what untold effort, above the chances, accidents, and irrationality of their animal origins.

The application of these criteria of value helps to explain or clarify many phenomena in the work of humanists and scientists. Perhaps we have felt dismayed by the shallow, even ugly, nature of much literature, painting, music, of the last several generations. Of course the contemporary often suffers in contrast with the past. The dross of the past has largely perished, or lies gathering dust in a museum or archive, which amounts to the same thing; and the best remains, against which to test the new. Also the contemporary is likely to depart from past patterns because the artist (as, too, the scientist) must keep seeking for new ways of communicating what comes from his continual reaching out into the unknown. This is not what I refer to, but to shallowness and ugliness intrinsic to the work.

At the same time we hear such work referred to as "scientific" or as being "influenced by science." This is also said of works other than artistic. For example, large numbers of people are persuaded, in the name of "science," to make use of some fruit of science when the results to be expected are essentially trivial, and the side effects scarcely known. It is in cases such as these that our criteria can help us to evaluate and explain, and to correct.

Early science is always analytical. You have to collect data, and describe and report phenomena, before you begin to see connections between them and patterns in the whole. A Linnaeus and a Mendeleev, the generalizers who recognize the largest patterns, come late in the development of any science. This is a state of affairs which leads many people to think of science as entirely analytical. For example, in the preface to a book on educational philosophy published in 1957, the author speaks of modern science as "no longer merely a report of particular facts," as though this was what one naturally expected science to be: a report of particular facts. If an educated person nowadays can make such a mistake, it is easy to see how in the past and to the present day some humanists, seeking to imitate science (because, perhaps of its manifest successes in coping with some aspects of the perceptual world) might make the mistake of imitating only the analytical part, not seeing the rest. Thus they would imitate *partial* science.

Out of such misunderstanding has come a spate of literature over the last seventy or eighty years: the reportorial, the minutely and solely descriptive, the stream-of-consciousness types of writing which are surely unsatisfying as art, though they may be good reporting. They are unsatisfying and misguided, on three counts. One is that they attempt to be scientific but imitate *partial* (analytical) science. Another is that they try to imitate science at all. We have seen that science and art are quite distinguishable in terms of their emphases on means of communication. The artist who attempts to imitate science runs the grave risk of losing his integrity as an artist. For if he were really successful, he would become a scientist, and no longer an artist. If he fails, then he has abdicated his duty to no valuable end. The third cause of failure is that they are largely analytical. This is why they are unsatisfying, and ugly. The great tradition of humanism confirms that humanists are concerned with wholes. No wonder reporting, by itself, seems superficial when masquerading as literature; why painting which is imitative without understanding and a theory is ugly.

Ignorance of a similar kind is found in technology, philosophy, and science. For example, it may be discovered that a certain spray will be extremely effective in killing mosquitoes or their larvae. Wholesale spraying, then, of parks, fields, streams, does indeed kill off the mosquitoes, but also the birds and fish which, had they been fostered, might have done a reasonably good job in the first place of killing the mosquitoes. (Solutions which pertain to be absolute—"to kill off *all* the mosquitoes"—should on principle be avoided, or in any event, thoroughly scrutinized.) This kind of error lies in mistaking analysis for all there is, or in looking at a complicated system with a very low-level-of-abstraction viewpoint, narrowly restricted. The proper functioning of analysis, synthesis, and reduction to practice would lead one to think of the whole situation. This is what ecologists are trained to do, and a well trained ecologist would not commit such mistakes.

There is a movement in philosophy (to give one more example) that is called "analytical philosophy." This seems to be the cul-de-sac into which logical positivism has worked itself. It may be a branch of linguistics. But insofar as it is largely analytical it cannot, in my view as stated above, aspire to the title of a philosophy.

These examples serve to suggest how our criteria of value may be applied. We ask of a work, whether of art or science or philosophy or technology, "does it fruitfully unite analysis, synthesis, and reduction to practice?" If it does, then it is greater, and more valuable than a work which does not.

This is not to deny value to the latter works, but we need to be able to compare, with reasonable assurance, the communications with which we are continually faced. We can then surely say that some potboiling experiment, or the collection of a series of properties of compounds, though having merit if well done, is less valuable than work containing elegant reasoning, or one which derives a relationship. We can look at a canvas on which has been splashed some handfuls of paint, and agree that it has some local merit as expressing the painter in his honest effort to devise an effective idiom, perhaps, but we would not rank it the equal of a canvas which carries to a group of artists a dream, or a message of hope, or power, or enlightenment.

What we are saying is that the application of these criteria, which are derived from a consideration of the whole compass of knowledge and experience as it is manifest in a university, enables us to evaluate in a reasonable way a great many of the utterances of our culture, whether they be works of the mind or the hand. Moreover, such evaluation can be objective and not imply condemnation of a work because it has merit only, and not great value. But we can demand of great art, and of great science, not only that it show us the way things are in this world—a good reporter can do that analytical job —but also that it show us how things could be. We ask that it elevate the spirit; that it lift us out of ourselves; that it set us an example of the heights to which the human being can ascend; that it show us how to ascend these heights.

We have endeavored in this essay to derive a few of the general principles which are common to, and so connect and unite, the arts and the sciences. We showed that there are authentic differences between science and art in terms of the means of communication used and the kinds of data amenable to the tools of artist and scientist. One cannot speak "artistically" about science when the communication is to be considered science any more than one can speak "scientifically" about art and have an artistic utterance. This is not a cause for dismay, nor is it a basis for conflict. The two are just different.

We have also showed that the artist and scientist use the same conceptual tools—analysis, synthesis, and reduction to practice—in performing their labors, and thus they meet at this common intellectual level. They are able, if they will look at function, to see these tools in operation, and not be misled by the myriad differences in appearance of the results. Thus they can understand each other at this level.

At the same time we are enabled to evaluate the works of the artist and scientist. The criterion of value we use is this, that that work in which the combined use of these tools produces a work which elevates the human spirit and shows us of what noble thoughts and acts we are capable, is greater than one which does not, or does less.

It is to such art and science that we can truly aspire. Our criteria allow us to recognize them even in areas outside of our specialities. We are thus fortified, in meeting our responsibilities as educated people, to speak to, to act out, and to teach, a passion for the first-rate.

BIOSYNTHESIS OF THE NICOTIANA ALKALOIDS

By R. F. Dawson*
Columbia University

Carl Wilhelm Scheele, in the latter half of the eighteenth century, isolated and described some of the earliest, known organic compounds other than sugar, namely, malic and citric acids from apple and gooseberry fruits and oxalic acid from rhubarb. Since Scheele's time many generations of organic chemists have sought and found novel constituents in plant materials. The flow of new discoveries long ago assumed major proportions; it continues unabated to the present day, and there is no indication of imminent exhaustion of supply. It may fairly be said that neither the animal kingdom nor the microbial kingdom contains so great a number and variety of chemical constituents.

Many compounds of plant origin have been assigned specific positions in that complex of consecutive biochemical reactions which is termed general metabolism. In addition,

* This research has been supported variously by the University of Missouri, Princeton and Columbia Universities, the Rockefeller Foundation, the U. S. Atomic Energy Commission at Brookhaven National Laboratory and at Columbia University, the Tobacco Industry Research Committee, and the American Tobacco Company.

many have been assigned specific functions—e.g. catalytic, structural, energy-yielding, information-storing—in this complex. For a vastly greater number, however, neither intermediary position nor metabolic function is known at present. This indeterminate group includes the terpenes, flavonoids, resin acids, glycosides, sapogenins, coumarins, alkaloids, and many others.

Organic chemical interest in the latter group, the so-called natural products, has extended well beyond the determination of molecular structure, total synthesis, and the correlation of structural parameters with pharmacological properties. The name of Sir Robert Robinson, especially, has been associated with efforts to infer probable biogenetic relationships between natural products and other more ubiquitous plant constituents [1]. Such inferences rest upon appropriate dissections of molecular models and upon arguments concerning the kinds of reactions that may underlie the structural analogies so revealed. The spectacular success obtained by extending the empiricisms thus derived to prediction of correct structure of relatively complicated molecules (e.g. emetine) has lent support to the possibility that these empiricisms may actually reflect modes of biosynthetic activity in the plant.

The challenging speculative possibilities raised by the organic chemist have not elicited an enthusiastic response from the botanist. The explanation probably lies in the fact that most natural products accumulate in the plant body in the manner expected of terminal metabolites or wastes and therefore appear to have little importance. Consequently, there has been very limited activity aimed at developing and characterizing plant systems which would be suited for the systematic investigation of natural-product biosynthesis.

However, the botanist whose scientific interests encompass molecular structure cannot fail to be impressed by the great numbers as well as the structural variety and potential complexity of these natural products. Further, he may note that the bulk of them occurs within the evolutionarily more advanced plant phyla, namely, the seed plants, ferns, and clubmosses. Finally, he may reason that, even though these substances may ultimately prove to be metabolic wastes, their occurrence in such great variety and at so late a point in plant evolution must signify something more than mere coincidence in the economy of nature.

In recent years limited progress has been made in placing questions of plant synthetic activity upon an experimental basis. Methods for the cultivation of plants and of plant parts under relatively controlled conditions have been developed. Many of these methods seem to be almost directly applicable to the development of systems for precursor testing. Likewise, isotopic tracers and methods for their application in biochemistry have become readily available. For a number of good reasons these methods have been employed principally in experimentation upon the biosynthesis of the tobacco alkaloids. Of all the so-called natural products, therefore, nicotine and its isomer anabasine are at present most fully understood. The events which have led to this situation possess some historical interest owing not only to the general importance of the tobacco plant and its alkaloids in science and industry but also to the fact that these events constitute in the aggregate a prototype for the investigation of natural products generally.

Our purpose is to outline the principal developments in the field of tobacco alkaloid biosynthesis, to speculate a bit upon their meaning and, in passing, to call attention to an instance of the critical necessity of isotopes for tracing pathways of intermediary metabolism. The author expresses his gratitude to Percy L. Julian, who, as friend and teacher, some 25 years ago provided the intellectual stimulus that led to much of the work described herein. To his collaborators, David R. Christman, Marie L. Solt, and A. P. Wolf, among others, the author wishes also to acknowledge a very substantial share in the accomplishments to be described.

The Tobacco Alkaloids

The tobacco or, more accurately, the Nicotiana alkaloids have a relatively simple molecular structure in which the pyridine ring is a common element. Attached to the pyridine ring in position 3, according to organic chemical notation, is another nitrogen-containing ring that may be 5- or 6-membered and that may contain no double bonds or as many as three. Further variation in this second ring may be attained by introduction of a methyl group on the nitrogen atom. At the point of attachment of the two rings is a carbon-carbon single bond. The most widely and abundantly distributed members of the group are nicotine (I), nornicotine (II), and anabasine (III). The last-named predominates in certain wild species of Nicotiana. Nicotine and, on occasion, nornicotine are the principal alkaloids of the usual commercial species and strains of Nicotiana.

Development of Experimental Systems

Early attempts to experiment with the biosynthesis of these alkaloids were handicapped by lack of knowledge concerning the relative capacities of the different plant organs for synthesis. Prior to 1940 it was assumed more or less tacitly that the green leaf is the center of chemical synthetic activity in the plant body. Experiments in which detached leaves were supplied with aqueous solutions of presumed precursors did not, however, yield definitive results. Consequently, a search for more suitable experimental systems was undertaken. A reading of N. P. Krenke's *Wundkompensation, Transplantation und Chimären bei Pflanzen* [2] had led to the suggestion that reciprocal grafts between tobacco and some related but alkaloid-poor species such as tomato might be used to indicate those portions of the plant body possessed of substantial capacity for nicotine production.

Preparation of such grafts and assay of their component organs yielded the surprising result that little nicotine was present in tobacco leaves grown on tomato roots, while tomato shoots grown on tobacco roots contained large amounts of nicotine. This observation, it ultimately developed, had been made before by workers in Japan, India, and Italy, but its meaning was not clear until, with accessory experimentation, we were able to show that nicotine is actually produced for the most part in the tobacco root system and transported in substantial quantity to the leaves where it accumulates. Similar conclusions were reached independently by Mothes and his students in Germany [3].

With attention thus focused on the root system of the tobacco plant, we turned to the methods for sterile culture of excised plant roots that had been described shortly before by White [4]. It was found possible to grow the young roots of Turkish tobacco (*Nicotiana tabacum* L.) and of certain cigar types (but not of many cigarette types) in a sterile aqueous solution of inorganic salts, sucrose, vitamin B_1, and yeast extract. Under standardized conditions, such cultures (Fig. 12) can be grown in an essentially steady state through an almost indefinite number of passages. The unit of transfer is a single growing root tip.

Three features of this system are most important for the present discussion. First, the growth rate of the isolated root, in terms both of dry weight and of total linear extent, follows

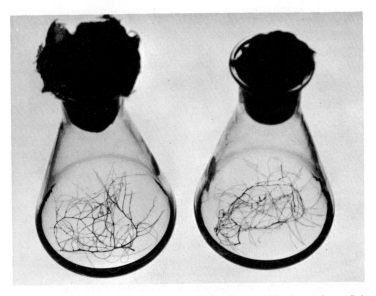

Fig. 12. Excised root cultures of Turkish tobacco (*Nicotiana tabacum* L.) in 125-ml Erlenmeyer flasks.

first-order kinetics for the first twenty-one days. Further, the growth rate constant, as measured by either criterion, is identical with that for nicotine production with proper attention, of course, to units (Fig. 13). However, the rate of root branching is also first order with a rate constant very nearly the same as that of growth or of nicotine production. It follows that the individual root branch grows and produces nicotine at constant rates which are simply proportional to one another. Now, the tips of both the primary root and its branches are known to be the loci of growth activity in the excised tobacco root (there is no active cambium in the excised root under conditions of culture). We have shown experimentally that the root tips are also the loci of nicotine formation. Clearly, therefore, one can regard the root tip in our experimental system as a uniformly moving point source of both nicotine and new root tissues. If the roots are subcultured regularly every three weeks, the growth and nicotine production rate constants remain relatively unchanged from passage to passage.

Parenthetically, it may be noted that excised root cultures of the wild species, *Nicotiana glauca* Grah., have also been obtained. These cultures produce relatively small quantities of nicotine, and the kinetics of such production parallel very closely in all but order of magnitude the situation described above. The same cultures produce anabasine as major alkaloid, however. Anabasine production follows entirely different

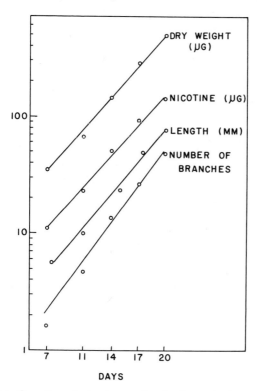

FIG. 13. Growth and nicotine production by excised Turkish tobacco roots during the first 20 days of culture. Figures on ordinate are 0.10 the true values for length and dry weight.

kinetics; under steady-state conditions, the alkaloid is produced in quantities that are approximately proportional to the dry weight of the root mass multiplied by the half-time of the culture in days. The latter relationship best fits a model in which alkaloid production occurs at a reasonably constant rate, not in the growing tips but in the mature tissues of the root organ. These close but contrasting relations between alkaloid production and growth status are deemed to be of special importance and will be reconsidered at a later point in the present discussion.

A second important feature of the sterile, excised root culture as an experimental system is the substantial and predictable nature of its alkaloid productivity. Under standardized conditions clonal material of Turkish tobacco roots produces as much as 0.2 mg of nicotine per culture in 21 days. In 28 days the yield may reach 0.4 mg per culture. Obviously, 20 cultures will yield sufficient alkaloid for ready isolation as the picrate, and numbers as large as 100–200 (in 125-ml Erlenmeyer flasks) can be managed with ease. The proportionality between nicotine and dry weight production figures is not easily altered by quantitative changes in nitrogen supply or by other simple environmental manipulations.

A third useful feature of the system is its microbial sterility. Organic compounds can be added to such cultures aseptically and their subsequent fate ascribed to the chemical activity of the tobacco root tissues alone.

With a first-class experimental system at hand, we may now consider what is to be accomplished by its use.

Biosynthetic Intermediates

Sir Robert Robinson, the eminent English student of natural product chemistry, long ago suggested that the pyridine ring may be related biogenetically to the diamino acid lysine (IV) and that the pyrrolidine ring may be derived

BIOSYNTHESIS OF NICOTIANA ALKALOIDS 125

similarly from a homolog of lysine, namely, ornithine (V). Both lysine and ornithine are, so far as is known, universally occurring metabolites. Trier, in his revision of Winterstein's *Die Alkaloide* [5], suggested that nicotinic acid (VI) may be a precursor of the pyridine ring of nicotine and that proline may yield the pyrrolidine ring. Nicotinic acid had not yet been isolated from plant materials, nor had it yet been shown to occur universally in cells as a component of the pyridine dehydrogenases. Therefore, Trier proposed that proline may act also as a precursor of nicotinic acid. Robinson criticized the nicotinic acid hypothesis on chemical grounds by pointing out the inertness of position 3 in the pyridine ring and the absence of a known case of displacement of the carboxyl group.

$$\begin{array}{ccc}
\text{COOH} & \text{COOH} & \\
| & | & \\
\text{CHNH}_2 & \text{CHNH}_2 & \\
| & | & \\
\text{CH}_2 & \text{CH}_2 & \\
| & | & \\
\text{CH}_2 & \text{CH}_2 & \\
| & | & \\
\text{CH}_2 & \text{CH}_2 & \\
| & | & \\
\text{CH}_2 & \text{NH}_2 & \\
| & & \\
\text{NH}_2 & & \\
\text{IV} & \text{V} & \text{VI}
\end{array}$$

(VI: pyridine ring with —COOH)

With Bothner-by we supplied lysine-ϵ-N^{15} as well as lysine-C^{14} to Turkish tobacco root cultures [6]. Neither substance contributed an appreciable amount of label to the nicotine that was formed during the culture period. The pyrrolidine ring contained most of the small amount of label that entered the molecule. Using excised root cultures of *Nicotiana glauca*, which produce about 4 parts of anabasine to 1 of nicotine, label from lysine passed readily into the piperidine ring of anabasine. Only a very small amount of label entered the nicotine produced by these cultures. It was concluded, therefore, that, while lysine is not the precursor of the pyridine ring of nicotine, it is a precursor of the piperidine ring of anabasine. This result confirmed in a microbially sterile system the results Leete obtained by supplying labeled lysine to the roots of the whole plants [7].

Leete, and Dewey et al. [8] independently demonstrated that ornithine-2-C^{14}, when supplied to the roots of whole tobacco plants, was taken up by the plants and incorporated into the pyrrolidine ring of nicotine. We have confirmed this result with the excised root system.

Robinson's hypotheses were validated, therefore, in the case of the ornithine-pyrrolidine ring relationship, but they had to be rejected insofar as the lysine-pyridine ring relationship is concerned.

Prior to the discovery of the sequence, ornithine → pyrrolidine ring of nicotine, we had prepared and utilized in our test system nicotinic acid-carboxyl-C^{14}. No appreciable quantity of radioactivity could be recovered in the alkaloid fraction. However, if ornithine were to be the precursor of the pyrrolidine ring, the carboxyl carbon of nicotinic acid would be eliminated in any event. To prepare ring-labeled nicotinic acid, we resorted to neutron bombardment, using the Brookhaven reactor. Labeling with C^{14} and with H^3 could thus be achieved by the following reactions: $N^{14}(n,p)C^{14}$, and $Li^6(n,\alpha)H^3$.

The acids so labeled were supplied to root cultures of the Turkish variety and of *N. glauca*. In all instances 8 to 9 per cent or more of the radioactivity supplied was recovered in the nicotine and the anabasine subsequently isolated. Specifically synthesized nicotinic acid-2-H^3 was incorporated into nicotine and anabasine in even higher radiochemical yields, and over 98 per cent of the label was recovered in position 2 of the pyridine ring by specific methods of degradation.

Despite the objections raised by Robinson, it is clear that the antipellagra vitamin and component of universally distributed biological dehydrogenating catalysts, nicotinic acid, may provide the pyridine rings of nicotine and anabasine. On balance, Robinson correctly predicted the precursor of the pyrrolidine ring; and so also did Trier, for proline and ornithine are metabolically interrelated, and labeled proline yields nicotine labeled in the pyrrolidine ring. Trier correctly guessed the precursor of the pyridine rings of these alkaloids, but his derivation of nicotinic acid from proline cannot now be accepted.

The objections of Robinson could be circumvented if it were to be shown, in his words, that the 3-substitution of pyridine in nicotine "indicates the site of an active position in a reduced pyridine precursor." Thus, if nicotinic acid were to undergo partial reduction, as it does in fact reversibly in positions 1 and 4 of the di- and triphosphopyridine nucleotides of the naturally occurring dehydrogenases, chemical union with the appropriate intermediate derived from either ornithine or lysine may be more readily conceived. This possibility was examined by preparing and supplying to the Turkish tobacco root cultures each of the four possible hydrogen isotope-ring-labeled nicotinic acids.

VII, VIII, IX, X

Label from all but position 6 was transferred to nicotine. It appears, therefore, that the hydrogen atom in position 6 of the pyridine ring of nicotine must come from some source other than nicotinic acid. Earlier experiments had shown that the carboxyl carbon of nicotinic acid is lost during the conversion of the latter to nicotine. The hydrogen atoms (and almost certainly the carbon atoms to which they are bonded) of positions 2, 4, and 5 are retained. Tso and Jeffrey have shown that the nitrogen atom is retained[9]. The carbon atoms at positions 3 and 6 are likely to be retained, since to postulate otherwise would require ring opening, a possibility for which there is at present no evidence. This virtually complete accounting for the atoms of nicotinic acid serves to focus attention on positions 3 and 6 and on the meaning of the losses of hydrogen and of carboxyl, respectively, during the conversion to nicotine (Fig. 14).

Loss of the hydrogen atom at position 6 could occur in at least two ways. First, the cell may oxidize nicotinic acid or a suitable derivative (e.g. the di- or triphosphonucleotide) to the corresponding 6-pyridone. A subsequent reduction step would restore the pyridine ring. Alternatively, the cell may reduce the nicotinic acid derivative to yield in effect a 3,6-dihydro intermediate. It seems quite likely that label would be lost from such an intermediate as a result of simple exchange for hydrogen of the aqueous environment. It is also possible that, if label were not lost in this manner, it could be lost at the required dehydrogenation step later in the reaction sequence as a consequence of an enzyme stereospecificity opposite to that of the initial reduction step. The researches of Vennesland afford ample precedent for such a possibility.

To test one of these alternatives, 6-hydroxynicotinic acid-N^{15} was prepared and supplied to the Turkish tobacco root cultures. No label was transferred to nicotine. Evidently, the second of the possibilities mentioned above must prevail, for other routes are not readily conceived. We hope to prepare labeled 1-glucose-1,6-dihydronicotinic acid and to resolve the issue by supplying this compound to our root cultures.

Meanwhile, it may be interesting to speculate upon the possible pathways and mechanisms involved in the reactions

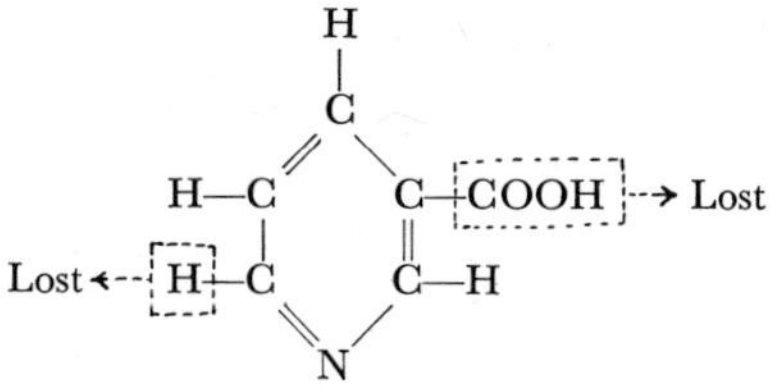

FIG. 14. The fate of the atoms in the molecule of nicotinic acid during conversion to nicotine. Unless otherwise indicated, retention is to be assumed.

of a reduced pyridine intermediate. Starting with a nicotinic acid derivative similar to one of the naturally occurring pyridine nucleotides, it is conceivable that this substance may become reduced 1,6 rather than 1,4 as in the case of the currently known codehydrogenases.

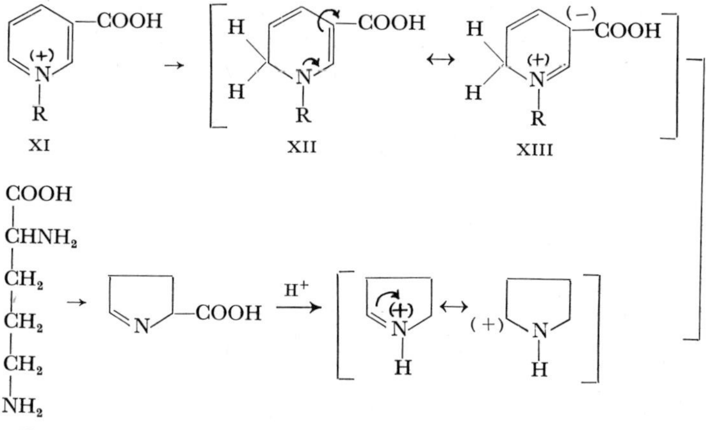

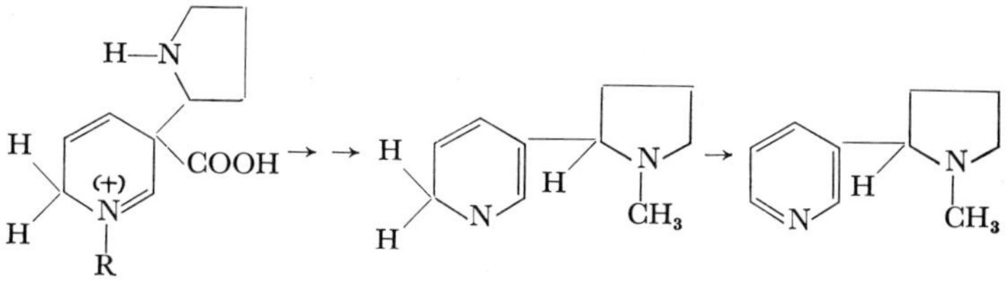

Alternatively, intermediate XII–XIII may be decarboxylated to yield 3,6-dihydropyridine (XXI). The latter may then undergo an aldol-type condensation with 4-aminobutanal derived from ornithine.

In both cases, the intermediate derived from ornithine reacts not with the relatively inactive pyridine ring but with a reduced pyridine precursor.

There is an apparently analogous situation in the biosynthesis of phenylalanine. Weiss and associates [10] have isolated an intermediate, prephenic acid, which may be regarded formally as a product of the addition 1,4 of pyruvic acid to p-hydroxybenzoic acid. Aromatization occurs enzymatically or in presence of traces of hydrogen ion by loss of carbon dioxide and water from positions 1 and 4 to yield phenylpyruvic acid.

In summary, it appears that nicotine and anabasine, respectively, may be synthesized in the plant from the universal metabolites, nicotinic acid and ornithine and nicotinic acid and lysine. Nicotinic acid may undergo some modification, probably a partial reduction, prior to reaction with the corresponding amino acid derivative. By this means the chemically difficult displacement of carboxyl may be explained. There is no indication in these experiments of the nature of the circumstances which compel the plant to synthesize apparently terminal products from such metabolically useful intermediates.

The Rate Stability of Alkaloid Production

One of the impressive aspects of alkaloid biosynthesis by the excised root cultures of *N. tabacum* and *N. glauca* is the extraordinary rate stability of the process. In our experience alkaloid production rate, in terms of dry matter production or of linear extension, is a clonal (i.e. inherited) character and is not subject to ready modification by the usual components of root culture environment. Of particular interest in this connection is the failure of added precursors such as nicotinic acid, either alone or with lysine or with ornithine, to increase significantly the yields of anabasine or of nicotine. Furthermore, it has not been possible to divert alkaloid synthesis by the excised roots of *N. glauca* even partially from nicotine to anabasine, or vice versa, by supplying lysine or ornithine in excess. In the case of *N. tabacum*, apparently up to 16 per cent of the molecules in the natural precursor pools may be substituted by similar molecules of extraneous origin. Aside from that, there seems to be no other effect even though as much as 50 per cent of the added nicotinic acid, for instance, may remain unused at the end of the experiment.

This situation stands in marked contrast to that in which excess nicotinic acid is fed to the chick. There, the acid is promptly combined with ornithine in the ratio of two moles of acid to one of ornithine and excreted as dinicotinoyl ornithine.

For reasons given earlier, the biosynthetic apparatus for nicotine and for anabasine are considered to be separated into two distinct compartments in the excised root, each of which is topographically and operationally definable, namely, the growing root tip and the matured root axis. This fact could possibly explain the observed difficulty in diverting biosynthetic activity from one alkaloid to the other by manipulating precursor supply, but the problem of explaining rate stability within compartments still remains. This task would be more easily accomplished if we knew the identity in each instance of the rate-limiting step and the detailed relationship of the latter to general metabolism.

The Possible Nature of the Rate-limiting Steps

Evidently the rate-limiting step for alkaloid biosynthesis in both *N. tabacum* and *N. glauca* lies somewhere between nicotinic acid and ornithine or lysine and the final product. The nature of this step or steps has not yet been ascertained, but some good possibilities already appear on the horizon.

The rate of formation of the postulated 1,6-dihydronicotinic acid intermediate would be, in the case of nicotine. formation at least, an attractive possibility, especially if the reductant were to originate with the reaction that is rate limiting in the over-all chemistry of the growth process.

There is a second possibility. Leete has shown that lysine is converted to anabasine through an asymmetric intermediate. The occurrence of a symmetrical intermediate between ornithine and the pyrrolidine ring of nicotine has been well established. The most plausible route to unsymmetrical intermediates lies obviously in the α-oxidative deamination of the diamino acid, whereas δ-oxidative deamination would lead to symmetrical intermediates.

$$
\begin{array}{c}
CH_2-CH_2 \\
| \quad\quad | \\
CH \quad *CHCOOH \\
| \quad\quad | \\
NH_2 \quad NH_2 \\
\text{XXV}
\end{array}
\rightarrow
\begin{array}{c}
CH_2-CH_2 \\
| \quad\quad | \\
CH \quad *CHCOOH \\
\| \quad\quad | \\
O \quad NH_2 \\
\text{XXVI}
\end{array}
\rightarrow
\underset{\text{XXVII}}{\left[\!\!\begin{array}{c}\\ N^*\end{array}\!\!\right]\!\!-COOH}
\rightarrow
\underset{\text{XXVIII}}{*\!\!\left[\!\!\begin{array}{c}\\ N\end{array}\!\!\right]\!\!*}
$$

$$
\searrow
\begin{array}{c}
CH_2-CH_2 \\
| \quad\quad | \\
CH_2 \quad *C-COOH \\
| \quad\quad \| \\
NH_2 \quad O \\
\text{XXIX}
\end{array}
\rightarrow
\underset{\text{XXX}}{\left[\!\!\begin{array}{c}\\ N^*\end{array}\!\!\right]\!\!-COOH}
\rightarrow
\underset{\text{XXXI}}{\left[\!\!\begin{array}{c}\\ N\end{array}\!\!\right]\!\!*}
$$

$$
\begin{array}{c}
\quad CH_2 \\
CH_2 \quad CH_2 \\
| \quad\quad | \\
CH_2 \quad *CHCOOH \\
| \quad\quad | \\
NH_2 \quad NH_2 \\
\text{XXXII}
\end{array}
\rightarrow
\begin{array}{c}
\quad CH_2 \\
CH_2 \quad CH_2 \\
| \quad\quad | \\
CH_2 \quad *C-COOH \\
| \quad\quad \| \\
NH_2 \quad O \\
\text{XXXIII}
\end{array}
\rightarrow
\underset{\text{XXXIV}}{\left[\!\!\begin{array}{c}\\ N^*\end{array}\!\!\right]\!\!-COOH}
\rightarrow
\underset{\text{XXXV}}{\left[\!\!\begin{array}{c}\\ N\end{array}\!\!\right]\!\!*}
$$

It is tempting to speculate that the root tissues that make nicotine may possess a predominantly δ-type of oxidative deamination of these diamino acids, while those that make anabasine may possess an α-type. Since the nicotine-producing compartment is transformed during growth into an anabasine-producing compartment, perhaps there is a concomitant shift in the specificities of the amino acid oxidases. Such a shift should not be difficult to detect. Specificity must extend beyond such elementary limits, however, for otherwise one would expect ornithine to yield unsymmetrically labeled nicotine in the matured tissues of the root at the same time that these tissues are producing anabasine.

The Dependency of Nicotine Production Rate upon Growth Rate

The close dependency of the rate of nicotine biosynthesis upon the rate of growth of excised Nicotiana roots in culture has been mentioned repeatedly. If the dependency were to be general and also obligatory, an entirely new aspect of alkaloid biochemistry and physiology would be brought to light. A successful analysis of this dependency might be expected not only to aid in answering the important question of the nature of the connection between alkaloid biosynthesis and general plant metabolism but also to provide some worthwhile and needed back doors into the complex and difficult chemistry of the processes that constitute growth.

A brief excursion into this field might begin with the inquiry whether nicotine production by plant organs other than the cultured, excised root is also closely linked to growth rate. We have found that nicotine production by the young, growing leaves of tobacco shoots grafted to tomato roots is linearly related to leaf dry weight[11]. However, the proportionality constants are very much lower than those relating nicotine production to dry weight accumulation in the root culture. The stem of the grafted tobacco shoot also makes small amounts of nicotine, and, indeed, it is the export of alkaloid from stem and graft union into leaf which ultimately disrupts the simple relation with leaf growth just mentioned. In the stem of *Nicotiana rustica* L., a linear relation occurs between length and nicotine content during the growth period. In the varieties of *N. tabacum*, however, nicotine yield is proportional to stem length multiplied by the half-time of the grafted shoot. These relations would be expected if alkaloid production were confined to the stem tip or bud of *N. rustica* but were to occur also in the matured regions of the two species of *N. tabacum*.

In all of these cases there is indication of a quantitative

connection between nicotine production and growth activity. Growth in stem length, for instance, occurs in and near the stem tip or apical bud, while growth in stem diameter occurs in an axial cylinder of tissues located below the bud and extending downward to the confluence with the root system.

Attempts to disrupt the rate dependency are still in their infancy but show some promise. For example, addition to the excised root cultures of β-indolylacetic acid (a generally accepted plant growth regulator) in 10^{-9} molar concentration resulted in a substantial reduction in growth rate and in a complete cessation of nicotine synthesis. Microscopic examination of the root tips indicated that the growth which had occurred was quite atypical; consequently, this effect deserves further investigation in terms of the kind of cell activity that may be correlated with nicotine formation. Gibberellic acid has been found not to alter appreciably the growth of roots in culture but rather to reduce their nicotine output by about 50 per cent. Mr. Robert Loe has found that 6-hydroxynicotinic acid may stimulate the nicotine output of whole plants relative to the rate of their growth. This last substance does not serve as a precursor of nicotine. Study of these effects is continuing and may ultimately yield useful tools for analyzing the rate dependency.

At present it seems well to adopt as a working hypothesis the prospect of a common rate-limiting step for both growth and nicotine production. This idea of commonality may also be extended to include the rate of anabasine production, if one could envisage retention in the matured root tissues of reaction pathways involved in an earlier growth phase. Since we know little of such pathways, the best we can do is to look for biological evidence. *Nicotiana glauca*, the prime producer of anabasine, is noted for its latent regenerative powers, especially when hybridized with *N. langsdorfii*. In our experience, the excised roots of *N. glauca* exhibit a strong tendency to regenerate buds through many passages following the initial isolation. Such behavior stands in sharp contrast to that of newly isolated cultures of the roots of *N. tabacum*. This tendency toward regeneration may represent a biological symptom of the survival in the adult plant body of a residuum of its special chemistry of growth. Whether or not such survival is related in any way to the capacity for anabasine production remains to be ascertained.

A Search for Botanical Correlations

Much of the preceding discussion has had as its purpose the creation, for the first time, of a rational topical outline into which the few known facts of alkaloid biosynthesis may be fitted. Our last problem is the nature of the relation between nicotine production and plant growth.

Growth in the plant may be divided into a number of somewhat less empirical components, empirical because they cannot be interpreted at the molecular level but are still recognizable with some precision by visual means. These components are, of course, cell division, cell enlargement, cell differentiation, and tissue and organ differentiation (cf. Wardlaw [12] for discussion). Each of these fundamental processes in morphogenesis occurs in the terminal centimeter of the tobacco rootlet where nicotine production also occurs. Each plays a demonstrable role in the growth of the root tip by the criteria that we have used for detecting growth. Each plays an equally essential role in the growth of the stems and leaves. The question at hand may be restated to ask whether the relative capacity for nicotine formation can be correlated with the occurrence of any of the visible manifestations of growth as listed above or with any of the physiological processes known to be ancillary to them.

Patterns of nicotine production in the intact plant have been fairly well delineated. The root system, as mentioned earlier, produces a disproportionate amount of alkaloid:

figures for the Turkish variety of *Nicotiana tabacum* are 97 per cent of the total plant alkaloid content produced in the root system, less than 1 per cent in the leaves, and the remainder produced in the stem. Obviously, the relative masses of root, stem, and leaf in a normal, untopped tobacco plant are not sufficiently different to permit the conclusion that nicotine formation is associated uniquely with cell division, cell enlargement, or cell differentiation in the general sense. Rather, if the process is to be correlated with a morphological entity or with a physiological process, the entity or the process must be distributed with extreme asymmetry between root and shoot organs. There are, of course, obvious differences in anatomy and morphology between root and shoot. Linear extension in the root system, for example, may be measured in miles, whereas it is measurable in feet in the shoot. Roots are probably the center of inorganic nitrogen assimilation by the plant. Further, they grow under conditions of generally lower oxygen partial pressure than do stems and leaves.

A few observations made upon plant parts or tissues cultivated apart from the intact organism are of interest in the present connection. Stem callus cultures are disorganized masses of tissue that are composed of rather large, highly vacuolated cells associated in a loose, friable mass. They are readily subcultured, provided proper nutrition is furnished. Relatively large quantities of such material (*N. glutinosa*) have been grown and examined for alkaloids. None could be found. Here, cell division, cell enlargement, and a minimum amount of cell differentiation have occurred; so also has nitrogen assimilation.

Through the kindness of Dr. Armin Braun of the Rockefeller Institute, we have been able to examine free cell cultures of a Turkish variety. These cultures are obtained by introducing stem callus tissues into a liquid medium. By continually shaking the culture flasks during the growth period, daughter cells are separated at the completion of division by the shearing action of the liquid. No alkaloids could be detected in this material. Since cell differentiation does not occur to an appreciable extent in this material, it is necessary only to point out that cell division, cell enlargement, and nitrogen assimilation, again, are not associated with nicotine production.

Dr. Braun also supplied cultures of a crown-gall teratoma culture. These cultures were obtained from bacteria-free crown-gall tumors on tobacco plants. The tumors were induced in stem tissue which possessed an unusually high regenerative power for buds and leaves. Small masses of such tumorous tissue were isolated from the plant by Dr. Braun and brought into sterile culture in the same manner as ordinary callus tissue [13]. This tissue, in sterile culture, retains a capacity to produce abnormal leaves and buds which have the external form but not the internal differentiation and organization of true stems and buds. Some vascular tissue is, however, produced in the basal portions of these tissue masses. The nicotine content of one lot of these teratoma cultures as supplied by Dr. Braun was 0.0006 per cent of fresh weight, a concentration far below that of even grafted leaves and stems of tobacco. This observation seems to remove cell division, enlargement, and limited differentiation from consideration even when accompanied by high regenerative power for aerial organs.

In the early stages of preparing one lot of callus cultures we observed that the freshly excised stem pieces placed on agar became covered with new, dark green growth [14]. The cells were small and very dense, but beyond that we know nothing of the histological and anatomical properties of the tissues so produced. In subsequent passages this appearance rapidly gave way to the typical callus culture appearance as described above. Nicotine assays yielded the picture given in Figure 15. The freshly isolated tissue mass produced a rather large amount of nicotine during its first passage in culture.

The quantities diminished rapidly in succeeding passages, however, and became undetectable when the typical callus state was attained. This case and that of the anomalous nicotine production of graft unions involving tobacco and tomato are the only known instances in which nicotine production rates of aerial tissues may approach that of roots. Nonetheless,

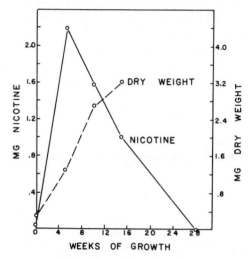

FIG. 15. The nicotine content of stem callus tissues of Turkish tobacco as a function of elapsed time in culture.

they suffice to negate the possibility that the higher nicotine production rate of the root is a consequence either of a difference in its environment or of an inherent root property. Rather, a connection with some histological or possibly organizational parameter is suggested. There are at present no further clues to the identity of this elusive parameter, but the entire subject now appears to be open for investigation.

Two final considerations are worth notice. In the genus Nicotiana many wild species are known, most or all of which contain relatively small amounts of nicotine and associated alkaloids. The cultivated species, *N. tabacum* and *N. rustica*, are not included among natural survivors, for they lost out in the struggle for survival long ago and have survived only through the efforts of man. It is an interesting fact that the cultivated species possess a relatively high alkaloid content, severalfold that of the wild species. It seems that alkaloid production is a genetic character which conferred so slight a survival handicap that the mechanism for its production is even now in the process of gradual elimination in the evolutionary continuum.

The distribution of these alkaloids among the different surviving groups of the plant kingdom is also worth notice. Nicotine is reported to be more widely distributed than any other alkaloid. It occurs in traces in the genera Equisetum and Lycopodium of the phylum Pteridophyta. Among the Spermatophytes or seed plants, nicotine or a related alkaloid occurs in the following species as indicated.

Order	Species	
Geraniales	*Erythroxylum coca*	(nicotine)
Rosales	*Sedum acre*	,,
Apocynales	*Asclepias syriaca*	,,
Polemoniales	*Nicotiana* species	,,
	Solanum melongena	,,
	Petunia violacea	,,
	Datura stramonium	,,
	Lycopersicum esculentum	,,
	Duboisia hopwoodii	(nornicotine)
	Atropa belladonna	(nicotine)
Asterales	*Eclipta alba*	,,
	Zinnia elegans	,,
Caryophyllales	*Anabasis aphylla*	(anabasine)

This distribution is widespread and involves orders that are considered to be relatively primitive as well as those that are more advanced in the evolutionary sense. It is entirely likely that further search with the sensitive methods now available may reveal the presence of nicotine, perhaps in very low concentrations, in still other orders among the Angiosperms. So far as present knowledge goes, however, the ability to make nicotine is confined to plants that possess leaves, stems, and roots and the vascular and other organized tissue systems to service such organs. These are the plants which became adapted, in the evolutionary sense, to grow on dry land. Once again, therefore, nicotine biosynthesis appears to be correlated with the development of tissues and of organs, although the correlation may not, and in fact does not, as we have seen above, extend to all tissues or to all organs equally.

Summary and Interpretations

The biosynthesis of nicotine from the universal metabolites, nicotinic acid and ornithine, has been described. The question was asked why it is that the plant would divert such potentially useful substances to the production of a compound which has all the earmarks of a waste product. A portion of the answer was provided when it was shown that the rate of nicotine biosynthesis is strictly dependent upon the rate of growth of the organ that produces the alkaloid. The probability is that the pathway leading to nicotine and the biochemical pathways of some phase of growth or development of the plant may share a common reaction or a common intermediate. This linking of alkaloid formation to such a fundamental process as growth or development constitutes a new departure in alkaloid physiology and, consequently, has been examined further in the hope of unearthing a correlation that might point to the general nature of the link.

The distribution of nicotine-synthesizing capacities among the different organs of the tobacco plant, the capacities of various excised organs and tissues in sterile culture to make nicotine, and the distribution of nicotine-synthesizing capacity in the plant kingdom were explored. Although no definitive correlation was found, it was possible to conclude that nicotine formation may have been a rather general property of land plants when these forms were first evolved; that the capacity has been suffering slow but definite attrition in the course of continued evolution; and that nicotine is probably formed during some phase of the histogenesis or of the organogenesis of the individual plant.

In chemical terms, nicotine has been regarded as an end-product of a primitive segment of the special chemistry of growth or development. The terminal steps of this segment have been considered to undergo kinetic depreciation in the long course of organic evolution. Quite possibly, further exploration of the pathways of Nicotiana alkaloid biosynthesis may afford a back door into an area of general biochemistry that, so far, has successfully resisted frontal assault.

REFERENCES

1. ROBINSON, R. *Structural Relations of Natural Products*, pp. 67–71. Oxford, Clarendon Press (1955).
2. KRENKE, N. P. *Wundkompensation, Transplantation und Chimären bei Pflanzen*. Berlin, J. Springer (1933).
3. MOTHES, K. Physiology of alkaloids. *Ann. Rev. Plant Physiol.*, **6**, 393 (1955).
4. WHITE, P. R. Cultivation of excised roots of dicotyledonous plants. *Am. J. Botany*, **25**, 348 (1938).
5. WINTERSTEIN, E., and G. TRIER. *Die Alkaloide*, 2d ed. Berlin, Borntraeger (1931).
6. BOTHNER-BY, A. A., R. F. DAWSON, and D. R. CHRISTMAN. Is lysine the source of the pyridine ring in nicotine? *Experientia*, **12**, 151 (1956).
7. LEETE, E. The biogenesis of nicotine. *Chemistry & Industry*, 537 (1955).
8. DEWEY, L. J., R. U. BYERRUM, and C. D. BALL. The biosynthesis of the pyrrolidine ring of nicotine. *Biochim. et Biophys. Acta*, **18**, 141 (1955).
9. TSO, T. C., and R. N. JEFFREY. Biochemical studies on tobacco alkaloids. I. The fate of labeled tobacco alkaloids supplied to Nicotiana plants. *Arch. Biochem. and Biophys.*, **80**, 46 (1959).
10. WEISS, U., C. GILVARG, E. S. MINGIOLI, and B. DAVIS. Aromatic biosynthesis. XI. Aromatization step in the synthesis of phenylalanine. *Science*, **119**, 774 (1954).

11. Dawson, R. F. Accumulation of nicotine in reciprocal grafts of tomato and tobacco. *Am. J. Botany*, **29**, 66 (1942).
12. Wardlaw, C. W. *Phylogeny and Morphogenesis*. London, Macmillan (1952).
13. Braun, A. C. A demonstration of the recovery of the crown-gall tumor cell with the use of complex tumors of single-cell origin. *Proc. Nat. Acad. Sci. U.S.*, **45**, 932 (1959).
14. Dawson, R. F. Nicotine synthesis in excised tobacco roots. *Am. J. Botany*, **29**, 813 (1942).

See also:

Anderson, R. C., E. Penna-Franca, and A. P. Wolf. *Brookhaven Nat. Lab. Quart. Progress Report*, October 1–December 31 (1954).

Dann, W. J., and J. W. Huff. Dinicotinoylornithine: A metabolite of nicotinamide in the chicken. *J. Biol. Chem.*, **168**, 121 (1947).

Dawson, R. F. An experimental analysis of alkaloid production in Nicotiana: The origin of nornicotine. *Am. J. Botany*, **32**, 416 (1945).

Dawson, R. F., and M. L. Solt. Estimated contributions of root and shoot to the nicotine content of the tobacco plant. *Plant Physiol.*, **34**, 656 (1959).

Dawson, R. F., D. R. Christman, A. D'Adamo, M. L. Solt, and A. P. Wolf. The biosynthesis of nicotine from isotopically labeled nicotinic acids. *J. Am. Chem. Soc.*, **82**, 2628 (1960).

Fisher, H. F., E. E. Conn, B. Vennesland, and F. H. Westheimer. The enzymic transfer of hydrogen. I. The reaction catalyzed by alcohol dehydrogenase. *J. Biol. Chem.*, **202**, 687 (1953).

Loewus, F. A., T. T. Tchen, and B. Vennesland. The enzymic transfer of hydrogen. III. The reaction catalyzed by malic dehydrogenase. *J. Biol. Chem.*, **212**, 787 (1955).

Rowland, F. S., and R. L. Wolfgang. Tritium-recoil labeling of organic compounds. *Nucleonics*, **14**, 58 (1956).

Solt, M. L. Nicotine production and growth of excised tobacco root cultures. *Plant Physiol.*, **32**, 480 (1957).

Solt, M. L. Nicotine production and growth of tobacco scions on tomato rootstocks. *Plant Physiol.*, **32**, 484 (1957).

Solt, M. L., and R. F. Dawson. Production, translocation and accumulation of alkaloids in tobacco scions grafted to tomato rootstocks. *Plant Physiol.*, **33**, 375 (1958).

Talalay, P., F. A. Loewus, and B. Vennesland. The enzymic transfer of hydrogen. IV. The reaction catalyzed by a β-hydroxysteroid dehydrogenase. *J. Biol. Chem.*, **212**, 801 (1955).

Wolf, A. P., C. S. Redvanly, and R. C. Anderson. The chemical consequences of the $N^{14}(n,p)C^{14}$ reaction in the acetamide system and the implications of nuclear recoil as a tool for synthesis. *J. Am. Chem. Soc.*, **79**, 3717 (1957).

CHROMOSOME REPRODUCTION AND THE PROBLEM OF CODING AND TRANSMITTING THE GENETIC HERITAGE

By J. Herbert Taylor
Columbia University

The Nature of Genetic Material

Chromosomes are small bodies, composed primarily of protein and nucleic acid, which are reproduced before a cell divides. Each of the two daughter cells resulting from a cell division usually inherits a complete set of the chromosomes characteristic of that cell. These in turn are accurately reproduced and each succeeding cell obtains a complete set. In higher organisms in which a body is composed of millions of cells, these are all descendents of a single original cell, the zygote or fertilized egg. This cell is formed by the union of a sperm and an egg. Each contributes a set of chromosomes that are similar but usually not identical. Indeed, the only contribution of the sperm in some species appears to be a set of chromosomes. Nevertheless, most hereditary traits are transmitted from generation to generation through the male line as well as the female line.

When the content of the sperm head—its nucleus or chromosome set—is examined chemically, only two substances are found, a class of nucleic acids called deoxyribonucleic acid (DNA) and a relatively inert type of protein [1]. In fish sperms the proteins belong to a class designated *protamines* which consist of about ninety per cent of the amino acid, arginine. The protamines are formed late in the maturation of the sperm and appear to be discarded soon after fertilization. Therefore, the only permanent component transmitted from generation to generation by the sperm appears to be DNA. This substance, nucleic acid, was discovered in cell nuclei back in the last century, but its relative inertness and apparent simplicity of structure made it an unlikely candidate for the transmission of genetic information. However, in 1944, Avery, MacLeod, and McCarty [2] showed that highly purified DNA isolated from one strain of bacteria could bring about a permanent genetic transformation of a related strain when taken up by the cells. Now attention began to focus on this substance as an important agent in recording and transmitting the genetic properties of cells. In 1952, Hershey and Chase [3] showed that the major component transmitted to the bacterial cell during viral infection was the DNA from the virus. More recently, several laboratories have shown that the reproduction of other viruses can be brought about by the transmission of only the nucleic acid component to the host cell. Viruses are composed almost exclusively of protein and nucleic acid, either DNA or the other type of nucleic acid, ribonucleic acid (RNA). From these experiments it is clear that the information for synthesizing both the proteins and the nucleic acids is recorded in the nucleic acid molecules.

Proteins are composed of long chains of amino acids laid down in precise patterns and folded in ways that are characteristic of each different type of protein, of which there are probably millions in the various species of plants and animals that exist on the earth. They are wonderfully versatile molecules, many of them able to catalyze chemical reaction that would be nearly nonexistent without these proteins or enzymes. Usually, each enzyme catalyzes one type of reaction and exhibits a high degree of specificity for the substrate or substances with which it reacts.

The protein's less versatile relatives, the nucleic acids, likewise are long chains of different units called nucleotides. The proteins may contain as many as 20 different kinds of amino acid, but the nucleic acids contain only four or five kinds of nucleotides. The nucleic acids have never been shown to have enzymatic properties. Indeed, they are quite unreactive, compared with the proteins. It is surprising, but perhaps because of this property, they have been selected during evolution of life to record and transmit the successful experiments of evolution. Their major role appears to be to transmit and perhaps to translate to the cellular machinery the information for making its proteins. The interactions of these proteins, the enzymes, with other molecules produce the multiplicity of substances found in cells and lead to the complex series of reactions which we designate cellular metabolism, or life, if you wish.

A Hypothesis—the Structure and Replication of DNA

In 1953 Watson and Crick [4] took the available chemical and physical information and constructed a scale model of the DNA molecule which has stood the test of time and serves to explain nearly all of the properties of this substance. The model consists of two chains of nucleotides twisted around each other like a winding staircase and cross-linked to each other by means of hydrogen bonds (Fig. 16). The nucleotides are units composed of a phosphate group, a five-carbon sugar (deoxyribose) and a purine or pyrimidine base. The polynucleotide chains are formed by a linkage of the phosphate

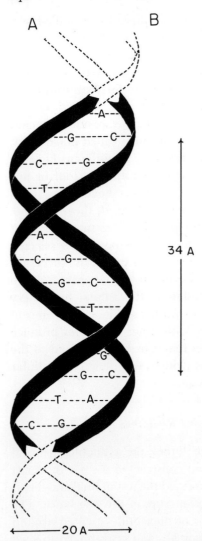

Fig. 16. Schematic model of a short segment of a molecule of DNA. The two polynucleotide chains are designated A and B. The purine and pyrimidine bases are represented by the letters A-T (adenine-thymine) and G-C (guanine-cytosine). See Figure 17 for details of hydrogen bonding.

groups and the sugars to form a long backbone of repeating units. The two backbones of a molecule differ only in a directional sense. The phosphodiester linkages extend from the third carbon atom in one sugar residue to the fifth carbon atom in the next sugar residue. Its third carbon in turn is linked to the fifth carbon in the next sugar. The two chains in the molecular model are reversed with respect to this 3–5 linkage. The variations along the chains are conferred by the four or five different purine and pyrimidine bases attached at right angles to the sugar residues of the backbone. The bases are paired, that is, hydrogen bonded, so that thymine forms two specific hydrogen bonds with adenine. Cytosine forms three hydrogen bonds with its partner guanine (Fig. 17). In addition, some molecules may contain various kinds

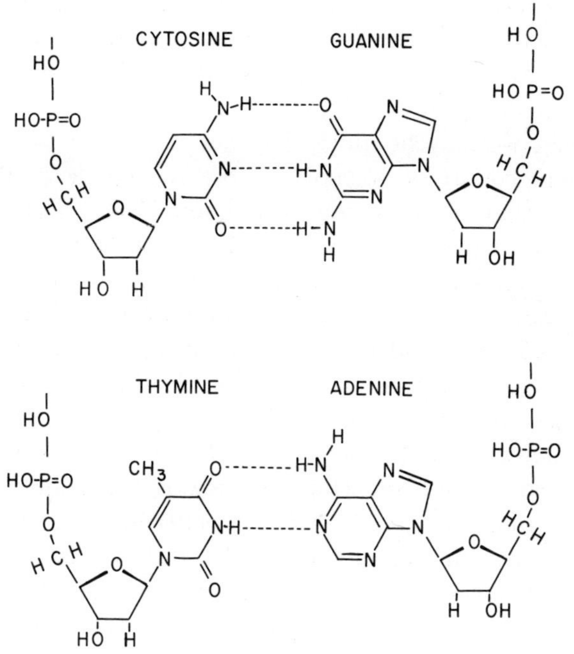

FIG. 17. The specificity of the two chains in the DNA molecule may reside in the hydrogen bonds between the pairs of bases. These base pairs lie in a plane at right angles to the long axis of the molecule and would represent the steps if the structure is compared with a winding staircase.

or derivatives of cytosine but all of these would specifically fit guanine by means of two or three hydrogen bonds. Hydrogen bonds are relatively weak bonds in which, for example, oxygen and nitrogen atoms share a common hydrogen atom when they are spatially located at the correct distance for interaction.

The molecule of DNA is relatively simple and the longitudinal variations consist essentially of the two units, pairs of nucleotides, that vary in their base pairs, A-T (adenine-thymine) or G-C (guanine-cytosine) (Fig. 17). Each nucleotide pair can be turned two ways in the winding staircase; therefore we can think of these long molecules as coded messages of four letters or characters. Several laboratories are now trying to understand the coding scheme. Ways and means appear to be available but the secret is still not revealed. However, we may be quite sure that these long molecules that may consist of 10,000 or more nucleotide pairs are sentences that will one day be read. Each DNA molecule may record part of the information for making a protein molecule, or possibly information for parts of several molecules. We are not yet able to define the genetic units, genes, in terms of the molecules. Indeed, the size and possible subunits of the molecule above the size of the nucleotide pairs are not yet well defined, but each chromosome appears to consist of many thousands of such molecules which we may think of as long taped messages on cellular metabolism. Certainly, each set of chromosomes could tell a wonderful tale extending back to the beginnings of life on earth if the decoding could be complete. Some of the chapters may have been lost, for many of the evolutionary experiments were unsuccessful and their records no longer exist in living form. Perhaps you have not looked at biological research in relation to the genetic code, but if you study morphology or taxonomy, you are reading—i.e. trying to put together and make sense out of—a part of this translated message. If you study physiology or ecology you read a different part of the message. If you study biochemistry you study still different parts of the message. In some aspects of the study (the properties of specific proteins, for example) one gets closer to the original code and the studies should begin to yield precise correlations, even with the techniques that are available to us now. Even if you study psychology you are studying some aspects of a remarkable and marvelous translation of the code into the mental patterns. Indeed, the tape appears to have created an organism with a mental capacity to wonder, to experiment, to discover, and eventually to decipher and perhaps to build a DNA coded according to his own specifications.

Although we cannot yet break the code, let us turn to the transmission of the coded information from one molecule to the two daughter molecules produced during reproduction or what we will call replication. Watson and Crick's model has the properties for faithful and accurate replication [5]. Each of the two chains characteristic of a double helix of DNA carries a pattern in the potential hydrogen bonds of its purine and pyrimidine bases for determining the sequence in a new chain when the two original ones are separated. If we refer to the chains as A and B (Fig. 16), we may think of one as a positive and the other as a negative in the photographic sense; they are complementary patterns or templates. When chain A is separated from chain B—and the necessary building materials, nucleotides, along with enzymes for coupling these by phosphodiester linkages, are available—two identical double helices, exact replicas of the original, could be formed. Chain A would serve as the template for a new chain B, and the original chain B would serve as a template for controlling the sequence of nucleotides in a new chain A attached to it in the form of a double helix. Here is a simple specific model and a testable hypothesis for its replication.

Given this hypothesis of replication, and the corollary hypothesis that each cell inherits a representative of each

DNA molecule, a precise scheme for the distribution of original and newly formed DNA chains at the cellular and chromosomal level would be predicted. One original chain of each molecule would be passed on to each chromosome and a new complementary chain would be built at each duplication. If the new chains could be labeled with radioactive atoms, they could perhaps be detected. After one or two additional duplications, their mode of transmission would serve as a test for the hypothesis of DNA replication. One would have to find a label selective for DNA and therefore chromosomes, a label that would remain in the molecules once they were labeled and, furthermore, a label that could be detected in objects as small as chromosomes. This is a large order, but as a result of the efforts of research workers in various laboratories all requirements could possibly be met.

An Experiment on Chromosome Duplication

Thymidine is utilized by some types of cells almost exclusively for DNA synthesis [6]. Once incorporated, the components of DNA molecules were shown to be highly stable; that is, exchange with other atoms or molecules in the environment was slow or nonexistent. By the technique of autoradiography, we were able to locate radioactive atoms in parts of a single cell. There appeared to be one way further to improve the resolution or precision with which the location could be accomplished. Tritium, radioactive hydrogen, emits a very low energy radiation when it decays. By placing photographic film in direct contact with a flattened cell and its chromosomes on slides [7], we could hope to decide which of these were labeled, that is, which contained radioactive hydrogen. The radiations would be stopped within less than a micron from their source and would register their presence by making visible silver salt crystals susceptible to development by the regular photographic techniques.

Tritium-labeled thymidine was prepared and supplied to plant roots in which cells were dividing [8]. After a time sufficient for the chromosomes of some of the cells to duplicate, the roots were washed to free them of the unbound radioactive molecules. Those not already utilized for synthesis of DNA were soon depleted and the labeled chromosomes arrived at division after a few hours. Chromosomes are visible in cells only during the short period of division. During the period before division the chromosomes could be shown to duplicate and become labeled when the tritium-thymidine was available to the cell. We prepared autoradiograms of the cells at division to see how their chromosomes were labeled. Indeed, each two daughter chromosomes descended from an original unlabeled chromosome was now labeled (Fig. 18, A and B). This would happen if each daughter chromosome inherited one-half of each unlabeled molecule and built a complementary labeled half along each. However, the same result could be obtained by a variety of other mechanisms of reproduction. The really important question to be answered was whether each chromosome also contained the original unlabeled DNA chains and, if so, would these separate from the labeled ones after one more duplication. We merely had to wait for some of the cells with labeled chromosomes to pass through another stage of duplication and arrive at the next division. Once we learned how long this took and how to recognize the cells at a second division, appropriate autoradiograms could be prepared. In the original experiments we placed the roots in a solution of a drug, colchicine, which stops movement of chromosomes into separate cells but does not prevent their separation and continued reproduction. Those cells with twice the usual number of chromosomes could be assumed to have gone through one, and only one, duplication after the drug was supplied. The chromosomes were labeled and then the cells were supplied with colchicine. When some cells had

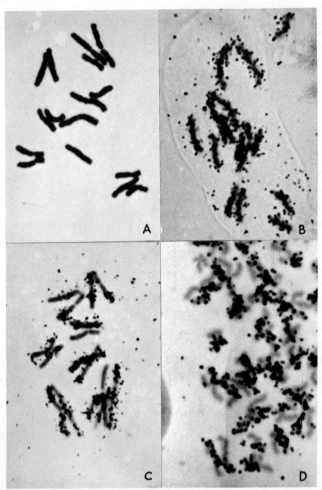

FIG. 18. A. The chromosomes in a root tip cell of Bellevalia blocked in mitosis by colchicine. ×1,500. B. Autoradiogram of a similar cell which had tritium-thymidine available during the duplication of the chromosomes which occurred several hours before when the cell was in the interphase (resting) stage. ×1,650. C. Autoradiogram of chromosomes in a cell which has resulted from the division of a cell similar to the one shown in Figure 18B. The chromosomes have duplicated once since they were labeled by tritium-thymidine, but the last duplication occurred in the absence of labeled thymidine. ×1,650. D. Autoradiogram of chromosomes in a human (HeLa) cell culture. See text. × Ca. 3,000.

accumulated twice the usual number of chromosomes, autoradiograms were again prepared. These showed what is apparent in Figure 18, c and d; that is, that one daughter chromosome was labeled and one was free of label, as the hypothesis had predicted. The cell shown in c has the usual number of chromosomes characteristic of the species, for after we learned the length of the division cycle we could find cells at the second division after labeling and therefore perform the experiment without the drug. This facilitates the making of good autoradiograms and eliminates the chance that colchicine would influence the result. Figure 18d shows a picture of human chromosomes from a cell grown in culture (strain HeLa). This one grew for about 30 minutes in a medium with thymidine-H^3 and was then transferred to a new medium until it had reached a second division after incorporation of the isotope. Those chromosomes having part of both chromatids (daughter chromosomes) labeled in c and d are the result of sister chromatid exchange during or after the labeling (see the following section). A small unlabeled chromosome visible in d may have resulted from the failure of all chromosomes to become labeled during the short contact with thymidine-H^3. Some species have an asynchronous duplication of the chromosomal complement and in these cases not all of the chromosomes will become labeled during a brief contact with tritium-thymidine [9].

The events of chromosome reproduction and the distribution of the radioactive subunits of the chromosome had been accurately predicted by the hypothesis. Such clear and straightforward results are rare in biological experiments, but the characteristic of a well-founded hypothesis is its ability to suggest experiments and to predict the results. In this sense, the hypothesis of Watson and Crick was elegant. However, we must recognize that few hypotheses can be proved by one experiment. The results predicted were found, but alternative explanations were possible. One fact appears clear, a chromosome consists of two subunits of DNA, one of which is regularly inherited by each two daughter chromosomes when the original duplicates, but the nature of the subunits is still obscure. One possibility was that the chromosome regularly carried two identical molecules of each type of DNA; i.e. the code in duplicate. In other words, the chromosome might be morphologically double and each half could in some manner produce an identical half which did not involve the separation of the two chains of each molecule of DNA.

Observation of the third division in broad bean roots and in Crepis root cells indicated that the number of labeled regions did not increase. In other words, no further longitudinal splitting of the DNA subunits could be detected. They were broken only along the long axis by the random exchanges which continued to occur at each succeeding duplication cycle.

Demonstration of a Difference in the Chromosomal Subunits

To learn more about the subunits of the chromosome, which we had discovered by the use of autoradiography, a more quantitative study was undertaken [10]. If the two subunits represent an array composed of one chain from each DNA molecule of the chromosome, they would not be identical. Like the two chains of the DNA molecule they would be complementary. One might suspect from the original experiments that they were not identical, for the four units present after duplication were distributed with great regularity—one old and one new to each daughter chromosome. An analysis of the pattern of exchanges between the two daughter chromosomes which were noticed in all of our original experiments promised to give some additional information. The pattern of exchanges could be determined in only the best cytological preparations and, of course, was visible only at the second division when the two daughter chromosomes (chromatids) are lying side by side, one labeled and one unlabeled. Exchanges, undetected until the second division, could presumably occur at any time during or after the original incorporation of tritium-thymidine. All events of the two duplication cycles would have to be deduced from the analysis of accumulated exchanges. First it is necessary to propose simple models of the subunits and to predict their behavior during exchange.

In Figure 19 (lower diagram) are given the results to be expected if the two subunits are different. The difference is represented as one of directional sense, for convenience and because this would be the difference between the two chains in a DNA molecule. Let us assume that exchanges usually involve both subunits in each chromosome; i.e. the two subunits of one daughter chromosome (chromatid) are reciprocally exchanged for the two in the other chromatid. If a difference exists, each new (labeled) subunit will be identical with the old one in the other sister chromatid. Let us assume that only like subunits can reunite. In that case, the labeled one will rejoin with the unlabeled subunit in each chromatid (Fig. 19). No visible change will be evident at the subsequent division if the exchange occurs during incorporation or before the first division after labeling occurs. Both chromatids will still be uniformly labeled from end to end.

Exchanges between only one of the subunits in a chromatid and one in the sister chromatid cannot be ruled out a priori. The expected results are shown in Figure 19 (upper diagram). A bridge might be expected to occur at the subsequent division. However, since such bridges are rare, this type of exchange does not appear to be making a significant contribution to the total exchanges observed and therefore will not be considered in our calculations.

Between the first and second division, exchanges should be possible during or after duplication. At this time an interchange would always exchange an unlabeled chromatid segment for a labeled one, because only one of the two chromatids would be labeled. All exchanges would be visible at the subsequent division. However, those occurring in the first and second interphases would be distinguishable if all of the descendants of the original labeled chromosomes could be examined. Those occurring at the first division would be visible in both descendants of the original chromosome, and

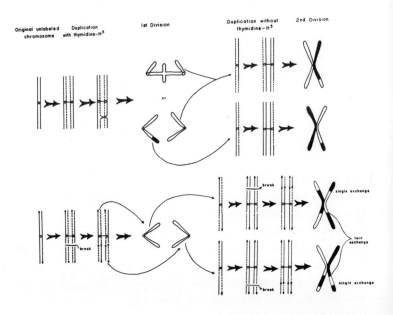

FIG. 19. A simple model showing the two subunits of each chromosome which were revealed by the experiments with tritium. The subunits are assumed to be unlike as indicated by arrows pointing in opposite directions. Exchanges that involve only one subunit in each of the two chromatids (upper diagram) appear to be rare or absent. Exchanges of the type shown in the lower diagram should result in a ratio of one twin pair of exchanges to each two single exchanges.

both exchanges would be at the same locus in the two chromosomes (chromatid pairs). In Figure 19 these have been referred to as twin exchanges. Those that occur in the second interphase are recognized as single exchanges. Now, if the cellular environments are the same at the first and second interphase, the number of singles should be two times the number of twin pairs—each is produced by a single reciprocal exchange and there are twice as many chromosomes at the second interphase when singles are produced.

If the subunits are identical the results in the first interphase would be different (Fig. 20). Of course, all exchanges in the second interphase would still produce singles. In the first interphase there are four possible combinations for rejoining two labeled and two unlabeled subunits—*1*) labeled and unlabeled subunits may rejoin in each chromatid; *2* and *3*) labeled and unlabeled subunits may rejoin in only one chromatid (two types); and *4*) the labeled subunits may rejoin with labeled ones in both chromatids. Only the first will produce a twin. The second and third produce one single exchange each, and the fourth produces no visible exchange. For each four exchanges in the first interphase, one twin and two singles should appear at the second division. For each of these four exchanges, eight additional singles should occur at the second interphase because there are two times as many chromosomes. Therefore, for each twin pair there should be a minimum of ten single exchanges. This is a minimum because, if the two subunits are alike, single exchanges should also be possible at any time before the second duplication by breakage of a chromatid, rotation of the two units by 180 degrees, and reunion of the break. This would rejoin a labeled and unlabeled subunit. The frequency of such events, if they occur, is not predictable in terms of the reciprocal exchanges, but they would increase only the number of single exchanges.

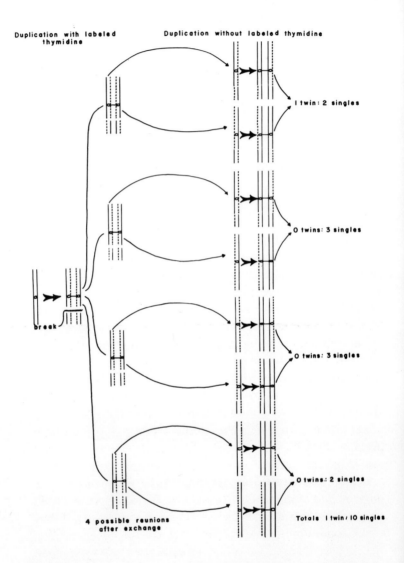

Fig. 20. Diagram showing the expected frequency of twin and single chromatid exchanges if the two subunits of each chromatid are identical.

TABLE 1. FREQUENCY OF SINGLE AND TWIN EXCHANGES IN THE TETRAPLOID CHROMOSOME SETS OF *Bellevalia romana**

Expt. Number	Number of Chromosomes Analyzed	Twin Exchanges	Single Exchanges	Frequency per Chromosome	
				First Interphase	Second Interphase
1	72 (18 cells)	36	15	1.00	0.21
2	112 (28 cells)	39	36	0.70	0.28
3	96 (24 cells)	14	26	0.29	0.27

* Data from Taylor [10].

The results obtained (Table 1) for the chromosomes of Bellevalia in the first experiment indicated that twin exchanges were more frequent than single [10]. This result is not predicted by the model unless more exchanges were occurring in the first interphase than in the second. If the subunits are different, all exchanges in the first interphase would produce twins. Therefore, we began considering possible factors that might be altering the number of exchanges in first and second interphases. Two differences were noted. *1)* Colchicine was applied to the cells after duplication of the chromosomes that would later be observed; therefore, the second duplication occurred in the presence of colchicine while the first did not. *2)* The amount of radioactive decay per chromosome was about twice as much in first interphase as in second. There was some indication from previous experiments that colchicine would reduce chromatid exchanges induced by radiation. Both factors, radiation and colchicine, could operate to increase twins.

The easiest factor to test turned out to be colchicine. Colchicine was applied to the roots along with the tritium-thymidine so that it was available in both interphases. Under these conditions the ratio shifted in the direction expected, if colchicine were reducing the chance for exchanges to occur. In addition, the total number in first interphase was reduced while the number in second interphase was slightly higher.

The colchicine appeared to be reducing the exchanges, but might be operating only after a delay. We therefore repeated the experiment with colchicine applied 2 hours before the thymidine. Now the ratio of twins to singles dropped to the predicted frequency and the total number of twins was reduced. The number of singles did not change.

The results clearly indicated that the two subunits of a chromosome are different. Unfortunately, the difference does not have to be one of directional sense. Furthermore, the difference would not necessarily be in the DNA molecules. If one assumes that protein chains run along the length of the chromosome the difference could possibly reside in these.

The Nature of the Chromosomal Subunits

Thus far, two facts concerning the chromosome had been established. *1*) Two subunits of DNA were present and these persisted in subsequent divisions except that exchanges continued to redistribute them slowly among the daughter chromosomes; no further longitudinal splitting of the subunits was evident. *2*) The two subunits were different in some way that prevented reunion of the unlike subunits. However, the nature of the subunits could not be defined, primarily because so little is known of the organization of the chromosome in the range between structures visible with the light microscope and the molecular level. Progress is being made at the level of resolution of the electron microscope, but at this writing a clear picture has not emerged.

Experiments performed on DNA replication in bacteria and also, later, on DNA replication in mammalian cells in culture appear to rule out the identification of our subunits with any morphologically visible subdivision of a chromosome.

Meselson and Stahl [11] performed the first one. They labeled the DNA with a heavy, nonradioactive isotope of nitrogen (nitrogen-15). Atmospheric nitrogen and organic nitrogen consist mostly of another isotope (nitrogen-14); DNA, containing these different isotopes of nitrogen, can be separated in a high-speed centrifuge in an appropriate salt solution. The two types of DNA, heavy and light, become layered in two bands in the centrifuge tube. Meselson and Stahl grew bacteria for many generations in nutrients labeled with heavy nitrogen. When essentially all DNA was heavy, the bacteria were transferred to a medium with nutrients containing the lighter isotope of nitrogen. Samples of the cells were then taken at intervals and the DNA particles (molecules) were released from the cells and brought into solution. At the time of change in medium, all of the DNA was of the heavy variety. The light variety did not appear immediately, as would happen if completely new molecules were formed. Half-heavy or "hybrid" molecules appeared first. Soon all of the original heavy molecules had disappeared and only the hybrid molecules existed in the cells. The hybrid molecules persisted, but after these had had time to reproduce they yielded one light molecule (unlabeled) and one hybrid (labeled) molecule, just as the labeled chromosomes had produced one labeled and one unlabeled daughter. These experiments have been repeated on human cells in culture with similar results, but with a different label [12].

The above experiments on the centrifugation of separated molecules or particles of DNA are further evidence that the two subunits of the chromosome share DNA chains in the same molecule and are therefore not identical half-chromosomes. However, the nature of the particles of DNA in solution in the centrifuge tube is still uncertain. Until they can definitely be shown to consist of Watson-Crick double helices, the mode of replication of DNA—that is, the means by which the genetic code is passed from cell to cell—cannot be considered finally established. However, since the experimental evidence was clearly predicted and the hypothesis explains and rationalizes so many results in genetics and biochemistry, it will be indeed surprising if the hypothesis does not hold up. Even as one experiment does not firmly establish a hypothesis, one contradictory result does not necessarily overthrow it; but, when any hypothesis no longer continues to explain and predict results, it has to be modified or perhaps discarded in favor of a better one.

The Synthesis of RNA in Intact Cells

Now let us come back to the matter of the translation of the genetic code in cellular metabolism. Although less is known and the mechanism appears to be more complex than the replication of the code, some intriguing ideas based on experimental evidence have been advanced. Since the DNA is usually restricted to the chromosomes in the nucleus of a cell but much of the metabolism, including the synthesis of many of the proteins, occurs in the cytoplasm, an intermediate molecule for carrying the coded information must be available. The other type of nucleic acid, RNA, is the most likely candidate. It occurs in both the nucleus and the cytoplasm. If it serves in this capacity, a part or perhaps all of it would have to be formed in the nucleus where the code would be transferred from DNA molecules to the RNA molecules formed with the DNA as a template.

The tritium-labeled nucleosides, cytidine and uridine, can be prepared and used to detect the synthesis of RNA in intact cells by the autoradiographic technique. When cells are placed in contact with these substances, the nucleosides are used almost exclusively for the synthesis of pyrimidine nucleotides and similar molecules, but they are not so selective for RNA as is the thymidine for DNA. Nevertheless, by taking advantage of our knowledge of the sequence of events in cells and the variations among cells, appropriate experiments can be designed to test the hypothesis that RNA should be formed in the nucleus. For example, we have recently used the cells of a rodent, the Chinese hamster, in tissue culture for such studies. These and other cells in culture are especially suitable for such experiments. Although we had demonstrated by autoradiography as far back as the early 1950's that the nucleus, and especially its nucleolus, was a very active site in the synthesis or accumulation of RNA, the recent experiments are more convincing and easier to illustrate.

When a cell is placed in a medium with tritium-labeled cytidine, the material quickly enters the cells, phosphate residues are added, and the nucleotides are incorporated into the RNA, presumably new RNA. After only 5 to 10 minutes the labeled RNA can be detected in the nucleus around the chromosomes and in the specialized bodies called nucleoli (Fig. 21A). None can be detected in the cell's cytoplasm yet. If the source of the label is now removed, the labeled, free nucleotides in the cell are soon used up or can be diluted by placing the cells in a medium with an excess of unlabeled cytidine and uridine. Cells fixed at intervals afterward reveal the disposition of the labeled RNA. In about 20 minutes after contact with the tritium-cytidine, labeled RNA begins to be detectable in the cytoplasm of the cell, and within 2 to 4 hours nearly all of the radioactivity can be shown to be in the cytoplasm, still in the form of RNA (Fig. 21B). The simplest interpretation is that all, or nearly all, of the RNA is formed in the nucleus, perhaps with the DNA as a template, and a large part of it then moves into the cytoplasm directly, or after some reorganization or addition of another component in the nucleolus. Actually, some of the RNA may be replicated in the nucleolus. From the recent experiments on the viruses, mentioned earlier, we know that RNA can code and transmit genetic information to other RNA molecules. Furthermore, it can cause infected cells to yield proteins specifically characteristic of the virus, even though only the viral RNA enters the cell [13,14].

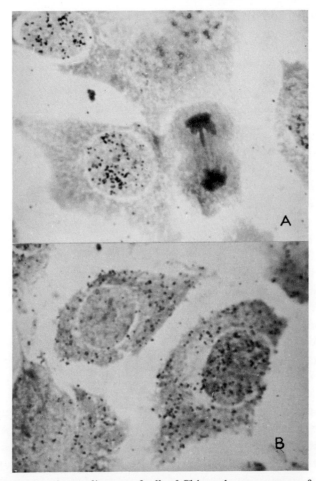

Fig. 21. A. Autoradiogram of cells of Chinese hamster, grown for 10 minutes in a culture containing tritium-cytidine. The grains are produced because the nuclei contain RNA into which the labeled cytidine has been incorporated. Note that cells in division stages do not form the labeled RNA. B. Autoradiogram of similar cells 4 hours after removal from the medium with tritium-cytidine. About one-half of the cells (upper left) have almost no tritium in the nucleus. The labeled RNA has moved to the cytoplasm. The other cells (lower right), in which chromosomes were duplicating during the period of contact with the labeled cytidine, retain tritium in the DNA of their nuclei.

To give some concept of the way in which nucleic acids may serve to guide the synthesis of specific proteins, I will briefly outline a hypothesis supported by recent experiments in biochemistry [15,16]. Nucleic acids do not appear to have a surface configuration that could directly orient amino acids in a specific way for linkage into the peptide chains of proteins. Their code more likely consists of the potential hydrogen bonds on the bases. RNA chains appear to exist as single unpaired chains, at least in part, in the ribosomes (small particles of the cell in which most of the protein synthesis occurs). In these ribosomes the RNA exists as large molecules but, in addition, in the soluble phase of cells much smaller molecules of RNA are found. These have been shown to attach specifically to amino acids in a reaction catalyzed by specific enzymes. There appears to be an enzyme, and a soluble RNA, which is specific for each of the 20 amino acids found in proteins. After attachment, the short segments of RNA serve as adaptors or handles which fit a specific site on the large RNA molecules. The specificity between the two RNA's, therefore, could reside in the specific hydrogen bonds of the bases. The large RNA molecules, properly oriented and with specific sites for the 20 adaptors, soluble RNA's, could then align the attached amino acids for coupling in a specific sequence characteristic of the ribosome RNA. The finished peptide chains should peel off, fold specifically because of the amino acid sequence or under the influence of some other part of the ribosome or associated machinery, and finally become a specific protein. These protein units may function directly as enzymes or, after aggregation, form the structural and enzymatic components of the cell.

I think we can begin to see in principle, at least, how the taped message in DNA can be translated in cellular metabolism and eventually form an organism. The link in the translation that still contains the largest gaps in the way of generalized hypotheses is the one that adequately explains differentiation of cells; i.e. how the code in one original cell, the fertilized egg, is made to vary precisely and sequentially in the construction of the many types of cells in such a complex organismal machine. Many problems of coding and translation of genetic information remain, but the next exciting breakthrough may be in this field of differentiation.

REFERENCES

1. FELIX, K. The nucleoprotamines: Their formation and their function. *Symposium on Molecular Biology*, R. E. Zirkle, Ed., pp. 163–177. Chicago, Univ. of Chicago Press (1959).
2. AVERY, O. T., C. M. MACLEOD, and M. MCCARTY. Studies on the chemical nature of the substance inducing transformation of pneumococcal types. Induction of transformation by a desoxyribonucleic acid fraction isolated from Pneumococcus Type III. *J. Exp. Med.*, **79,** 137 (1944).
3. HERSHEY, A. D., and M. C. CHASE. Independent functions of viral protein and nucleic acid in growth of bacteriophage. *J. Gen. Physiol.*, **36,** 39 (1952).
4. WATSON, J. D., and F. H. C. CRICK. Molecular structure of nucleic acids. A structure for deoxyribose nucleic acid. *Nature*, **171,** 737 (1953).
5. WATSON, J. D., and F. H. C. CRICK. Genetic implications of the structure of deoxyribonucleic acid. *Nature*, **171,** 964 (1953).
6. FRIEDKIN, M., D. TILSON, and D. ROBERTS. Studies of deoxyribonucleic acid biosynthesis in embryonic tissues with thymidine-C^{14}. *J. Biol. Chem.*, **220,** 627 (1956).
7. TAYLOR, J. H. Autoradiography at the cellular level, in *Physical Techniques in Biological Research*, G. Oster and A. W. Pollister, Eds., vol. 3, pp. 545–576. New York, Academic Press (1956).
8. TAYLOR, J. H., P. S. WOODS, and W. L. HUGHES. The organization and duplication of chromosomes as revealed by autoradiographic studies using tritium-labeled thymidine. *Proc. Nat. Acad. Sci. U.S.*, **43,** 122 (1957).
9. TAYLOR, J. H. Asynchronous duplication of chromosomes in cultured cells of Chinese hamster. *J. Biochem. Biophys. Cytal.*, **7,** 455 (1960).
10. TAYLOR, J. H. The organization and duplication of genetic material. *Proc. 10th Intern. Congr. Genet.*, **1,** 63 (1959).
11. MESELSON, M., and F. H. STAHL. The replication of DNA in *Escherichia coli*. *Proc. Nat. Acad. Sci. U.S.*, **44,** 671 (1958).

12. Simon, E. H. The transfer of DNA from parent to progency in a tissue culture line of human carcinoma (strain HeLa). *J. Mol. Biol.*, **3,** 101 (1961).
13. Frankel-Conrat, H. The role of the nucleic acid in the reconstitution of active tobacco mosaic virus (Communications to the Editor). *J. Am. Chem. Soc.*, **78,** 882 (1956).
14. Gierer, A., and G. Schramm. Infectivity of ribonucleic acid from tobacco mosaic virus. *Nature*, **177,** 702 (1956).
15. Crick, F. H. C. On protein synthesis. *Symposia Soc. Exp. Biol.*, **12,** 138 (1958).
16. Hoagland, M. B. Nucleic acids and proteins. *Sci. American*, **201,** 55 (1959).

SEX REVERSAL IN ANIMALS AND IN MAN

By EMIL WITSCHI
State University of Iowa

Sex and Its Accessories

IN COLLOQUIAL SPEECH "sex" denotes psychologic and morphologic sex characteristics that immediately come to mind; it is not generally known that these are secondarily acquired features of fancy or adaptation, which, in the course of evolution, were added to *basic sex*. It is common knowledge to all who have had so much as a freshman course in biology that many protozoans and algae produce only one type of germ cell, called the *isogamete*. At no stage of their life cycle does female or male differentiation occur. Nevertheless, it cannot be doubted that such species too are reproducing *sexually*. The fact that in fertilization *two* gametes always combine to form a zygote might indicate that some sort of physiologic or genetic sex differences exists, even at the lowest levels. In many instances the isogametes cannot definitely be labeled *male* or *female*, since they can act as one or the other "sex," depending upon the kind of partner they meet with. Kniep [1] and Hartmann [2], who have studied this situation most extensively, apply the expression "relative sexuality"—an appropriate term which, however, does not contribute further to the understanding of sexuality.

The most general characteristic of sexually reproducing organisms from bacterium to man is indeed not male-female differentiation, but the *cyclic change of haplophase and diplophase*. As a consequence of fertilization the chromosomes are doubled in number, and in the course of the meiotic divisions, single sets are again restored. The two phases may be similar

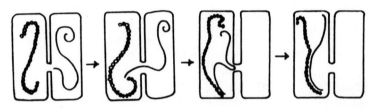

Fig. 22. Chromosome conjugation after fertilization in bacteria; after Lederberg [3].

in length and in morphologic appearance, as in many foraminifers and algae. More often one phase is more extended and developmentally elaborate than the other. In the search for the very essence of sexuality, *chromosome conjugation* emerges as the most significant and yet unfathomable phenomenon (Figs. 22, 23). If anywhere in nature, here one might be tempted to speak of mystery. The process is at once the ultimate consequence of fertilization and the preparation for meiosis. Allelic elements of chromosomes, after some sort of "mutual recognition," approach each other and, after contact and possible interaction, separate again with amazing, orderly precision. Obviously, a study of the mechanics and chemistry of these operations would bear on the essence—lead to the very kernels of the sex problem—but research in this field lags, awaiting the development of new methods of investigation.

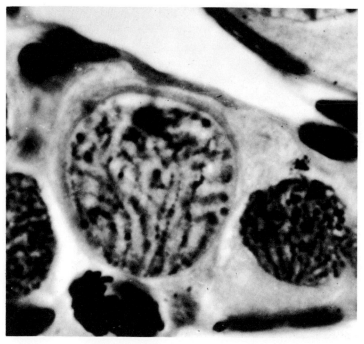

FIG. 23. Chromosome conjugation in an amphibian ovocyte [4].

Evolution of Sex

The present paper is concerned with the far less fundamental but much better known *accessory sex characters*, which were acquired in the course of evolution, often in adaptation to special living conditions. Separation into large macrogametes and small microgametes is widespread even in some free-living protozoans. However, the formation of typical eggs with large food depots and of extremely small, often highly motile sperms is correlated with the evolution of multicellular organization (Fig. 24) and with parasitism (e.g. Coccidia). The new acquisitions do not change the cyclic sex processes but may aid their perpetuation.

In multicellular organisms separation of somatic from the germinal cell strains provides almost unlimited possibilities of sexual diversification. Starting with duct systems that carry eggs and sperms, somatic female and male differentiation becomes more and more elaborate. Biologically speaking, mother love, and even sentiments as delicate as the yearnings of poem-writing youths are, like the fanciful plumage of a male paradise bird, mere subsidiaries to basic sex. Not every feature of this evolution must contribute to the maintenance of the basic sex cycles and the accomplishment of chromosome conjugation, but all must at least permit their existence and perpetuation.

In comparison, relatively little progress in morphologic differentiation was achieved by the germ plasm. Even in man the general appearance and the life cycle of the germ cells are still similar to those of parasitic endamebas and coccidia. Shortly after segregation from the soma cells, the diploid protogonia leave their original location in the endoderm and crawl through the mesentery toward the gonadal sites (Fig. 24A). Here they become temporarily encapsulated by somatic nurse cells until they are released again, many years later, as mature germ cells, haploid eggs, and sperms (Fig. 24B).

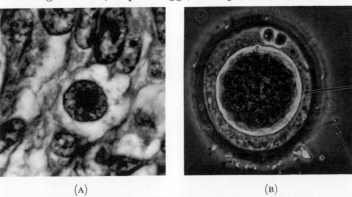

(A) (B)

Fig. 24. Human germ cells. (A) Primordial germ cell [5]; (B) mature egg and sperms. (After Shettles [6], from Witschi [4].)

One may illustrate the progress of the evolution of man graphically by a plumed serpent (Fig. 25), its tail reaching back into the dim past, the head projecting into the evolutionary future. One might think that the protozoan time, one to two billion years ago, was the epoch of the origin of life. In

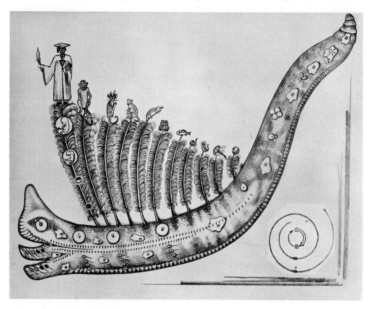

Fig. 25. Plumed serpent. The body represents the evolution of the human germ plasm. The plumes indicate evolution of somatic structure on the basis of mutational changes in the germ plasm. An allegorical diagram.

fact, some of the most essential elements of organismic life must have evolved even earlier, such as protoplasm and chromatin, the patterns of anabolic and catabolic metabolism with Krebs cycles, sensitivity and contractility, reproduction, and eventually also sexuality. The primitive precursors of organisms gained only gradually the cellular level. At later stages, sexuality, with fusion of entire adult cells or of their essential contents became established. Primitive fusion in turn was replaced by copulation of gametes. Isogamy, anisogamy, and ovogamy followed each other, the last probably achieved more than half a billion years ago, when the evolution of multicellular *metazoa* was under way. The next decisive progress, the evolution of *genetic mechanisms of sex determination*, can be located with some accuracy in the triassic and jurassic periods after separation of the saurian from the mammalian classes [7]. Originally these mechanisms certainly were based on genic mutations, affecting only a few loci. In the course of time, for reasons still entirely obscure, progressive changes spread over the chromosomes, resulting in the phenomenon of sex-linked inheritance and in microscopically recognizable *sex chromosomes*.

Thus man has reached a stage where actually three types of mature germ cells are produced: eggs with an X chromosome, gynosperms with an X chromosome, and androsperms with a Y chromosome. Since there is no reason to assume that, with present conditions evolution has attained a final stage, the continuation of progressive differentiation logically leads us to expect as the next step a complete separation of the male line into gynosperm-producing and androsperm-producing individuals. This forecast of a race with two types of men bears with it the promise of a very simple answer to the age-old desire for control of the sex of offspring. Indeed, mutations affecting the survival or even the maturation of either the female-determining or the male-determining sperm type might possibly occur any time now. In the meanwhile, efforts to separate X- and Y-carrying sperms by laboratory technics are being made and continued in spite of great difficulties and, so far, discouraging results.

Until geologically recent times, the evolution of mankind depended entirely upon mutational progress of its germ plasm. Some distant unicellular ancestors may have lived in shells of calcareous or other materials during certain phases of their life cycle. However, with the coming of the metazoan era the housing situation assumed entirely new aspects. Although somatic differentiation also rests on the genetic constitution carried by the potentially immortal germ plasm, the soma itself has no permanency of its own. The bodies perish with the change of generations, like the feathers of a bird that are cast off during molting periods. The often discussed question of whether man might have descended from embryonic rather than from full-grown stages of ancestral forms is utterly meaningless, since his predecessors were never other than ameba-like germ cells, the sole links between generations. The plumes of the symbolic serpent represent not only the temporary nature of the soma; they also serve as the indicators of mutational progress in genic constitution of the germ plasm, at invervals amounting to hundred-millions of years. With the advent of the scholar and the teacher, a fundamental change has occurred, inasmuch as now an ever increasing amount of information is passed on to subsequent generations over new channels, independent of the germinal pathway.

Sex Determination

a) Induction. Turning now to a consideration of sex determination—that is, of the circumstances which decide whether a primitive germ cell should become an egg or a sperm—it is soon obvious that in primitive and hermaphrodite forms *sex is induced* by the inner or the external environment. Nutritional factors play an important role. For lower organisms the botanist Goebel [8] generalized his impressions by saying that male germ cells are formed under conditions that are not sufficient for the differentiation of eggs. However, quantity is not always so decisive as quality. Even in the marine worm Ophryotrocha (Fig. 26), where small individuals always are males, and starvation changes females into males, Hartmann [2] finds evidence also of qualitative differences among various

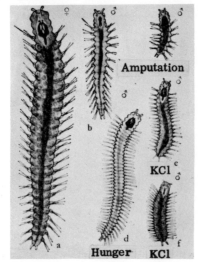

Fig. 26. *Ophryotrocha puerilis*, a polychete marine worm. Sexual differentiation depends upon external and internal environmental conditions; after Hartmann [2].

agents. Hormone-like substances have been extracted by Baltzer from adult females of Bonellia, which masculinize free-swimming larvae [9]. Similar masculinizing substances are produced by females of many other species, even some fishes. In these forms the alternative of attachment to a female or of free life becomes a sex-determining mechanism.

In *hermaphrodites* that produce eggs and sperms side by side, induction becomes localized in separate territories and often issues from particular cell types. In the snail Valvata the sex gland starts from an undifferentiated apex. Descending through the upper coils of the shell, eggs develop in a cortical layer, sperms in the medulla (Fig. 27a). Each ovocyte is provided with a follicle of numerous nurse cells, while many spermatocyctes have to share a common nurse cell [10]. The undifferentiated sex glands of all tetrapode vertebrates have a similar topography (Fig. 27b). When they differentiate later into testes, the cortex is reduced, while in the differentiation of ovaries the medulla becomes vestigial.

b) Genetic sex determination. The snails and vertebrates just considered differ essentially in that only the snails remain

SEX REVERSAL IN ANIMALS AND MAN

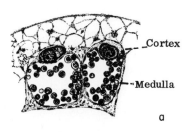 

FIG. 27. Inductive sex determination: The cortical territory induces ovogenesis, the medullary territory spermatogenesis. (a) Part of hermaphrodite sex gland of the snail Valvata; after Furrow [10]; (b) cross section through sexually still indifferent sex gland of a vertebrate (amphibian) with cortical and medullary primordia.

lifelong hermaphrodites. Amphibians, birds, and mammals become either males or females at an early, embryonic stage; they are *gonochorists*, i.e. animals with separated sexes. Moreover, the decision about which sex should dominate is primarily decided by *genetic mechanisms*. It is now well-known that in birds the female is heterozygous; the hen has only a single sex chromosome (ZO) while the cock has two (ZZ). In mammals and in man, it is the male that is heterozygous and usually has an unequal pair of sex chromosomes (XY). In principle the situation is much as in the Drosophila fly (Fig. 28), where sex is decided by a balance between male-

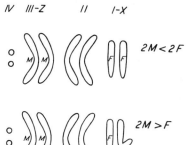

FIG. 28. Chromosomes of *Drosophila melanogaster* with indication of the localization of the sex-determining male (M) and female (F) gene complexes.

determining factors in the third and female-determining factors in the X chromosomes [12,13]. In a few Drosophila species the Y chromosome has disappeared and the male flies are of the XO-type, analogous to the ZO-type of hens.

There is often a tendency—even among good geneticists, but particularly by those that are insufficiently familiar with actual facts—to attribute almost mythical and absolute powers to chromosomes. Genetic sex determination would

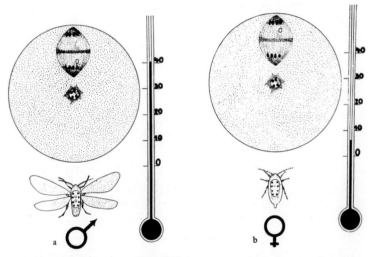

FIG. 29. Diagram by Ankel [15] illustrating the influence of extreme temperature on the direction of the movement of the Z chromosome in a moth; experiment by Seiler [14].

then be wholly different from determination by environmental induction. In a very instructive experiment, however, Seiler [14] has shown that external factors may gain control not only of sex but even of the sex chromosomes (Fig. 29). As in birds, the female moth *Talaeporia tubulosa* has a single Z chromosome; it may be pushed into the polar body or be retained in the egg, according to temperature conditions, which thus determine the sex.

c) *Postgenetic sex reversal.* An even more drastic alteration of the course of genic sex determination occurs in the case of the ovariotomized hen, which becomes a cock. Benoit[16] first showed that this is a consequence of testicular development of the right-side rudimentary sex gland. This example is of great theoretical importance in two respects. It furnishes

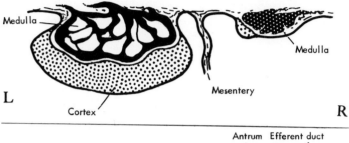

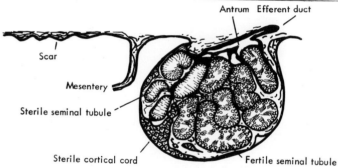

Fig. 30. Sex reversal in the hen by precocious ovariotomy. Upper figure: composition of sex glands in female chick at hatching. Lower figure: effect on development about one year later.

proof that the germ cells do not differentiate in response to genetic constitution, but to *inductive environment*. Because of the peculiar anatomy of avian sex glands, removal of the single ovary eliminates practically the entire cortical inductor. This permits the compensatory development of the medullary rudiment of the right side; it becomes a testis, secreting male hormones (growth of the comb!) and producing sperms (Fig. 30). The entire spermatogenesis is accomplished without a change in the female type (ZO) chromosomal constitution (Fig. 31)—a most important fact.

Similar sex reversals have now been accomplished in several vertebrate species and by a great variety of experimental methods. All have in common that inversion is

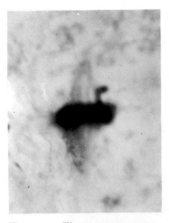

FIG. 31. First spermatocyte division in the "testis" of an ovariotomized hen. The single Z chromosome moves to the upper pole [11].

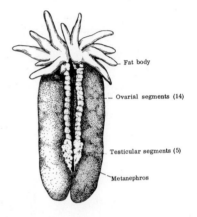

FIG. 32. Sex glands of a genetic male Xenopus, treated with estradiol during the first two days of the sex-differentiating period [17].

brought about by reduction and suppression of the dominant inductor system, followed by recrudescence of its constitutionally weaker alternate. Genes produce their effects by *series of postgenetic reactions*, which eventually determine whether cortical or medullary induction should prevail. These postgenetic events are open to natural or experimental interference. The most complete success in such managed sex inversion we have had so far is with the South African toad Xenopus. Here, hormone applications take the place of Benoit's knife in the suppression of the dominant inductor. At first our experiments met only with partial success. Female larvae treated with androgenic hormones, such as testosterone, acquired the secondary sex characters of males, including male sex behavior, but retained essentially normal ovaries. Full success came, however, when genetically male larvae were treated with female sex hormones. As little as 50 micrograms of estradoil per liter of water changes differentiating larvae within 3 days into females [17]. If treatment is interrupted after the second day, only a hermaphrodite is obtained (Fig. 32). If given the entire treatment, the genetic male is changed into a female for the duration of its life. Instincts and behavior are those of the regular female and, mating with a normal unchanged brother, it produces thousands of fertile eggs (Fig. 33), all of which, after insemination,

FIG. 33. Xenopus. Copula of a normal male (ZZ) and a sex-reversed male (ZZ) producing a batch of all male eggs.

are of male (ZZ) constitution. This experiment of breeding males (ZZ) with sex reversed males (ZZ) has now been carried on into the fourth generation, with complete exclusion of the heterozygous sex (ZW). It is an interesting consideration that an analogous experiment with the human species, if ever successful, might lead to the extinction of the old-fashioned heterozygous (XY) male.

d) Sex reversal by blastophthoria. Of the many other means of producing sex reversal, the method of delayed fertilization and consequent overripeness of the egg is of particular interest. Such damage to the egg can produce a whole series of malformations and one individual is often simultaneously afflicted with several of them [18]. For instance, a toad of female genetic sex was a hermaphrodite, obviously arisen from an abnormal egg, since it had an accessory small pelvis with stubby legs. It produced, acting as a father, over 2,000 offspring [19]. Systematic experiments on delayed fertilization of frog eggs result in the production of a large number and a considerable variety of malformations in embryos, larvae, and frogs. Impairment of the gastrulation process leads to defective head formation. Sometimes only rostral parts, face, hypophysis, and mouth parts are involved. More grave malformations are microphthalmia, microcephaly, and acephaly. Edema and malformations of the limbs—duplications as well as reductions—are frequent occurrences.

FIG. 34. Diagrammatic cross sections through gonads of frog larvae before sexual differentiation. (A) normal ratio between cortex and medulla; (B) and (C) rudimentary condition (two grades) of the cortex as brought about by overripeness of the egg; after Witschi et al. [21].

Particularly sensitive to overripeness are the germ cells, number and quality are far below normal standards as early as the tailbud stages. Migration to the gonadal sites is slow and irregular. This situation results necessarily in an underdevelopment of the gonadal cortex and compensatory enlargement of the medulla. Differences between genetic males and females never find full expression. In the end, some gonads may be entirely depleted of germ cells and remain small and sterile, while others become testes of low fertility (Fig. 34). Genetic females thus may be completely masculinized [20, 21].

Sex Deviations in Man

Abnormalities of sexual development occur with higher frequency in man than in most other mammals. Five characteristic types can be recognized, although almost every case has its individual peculiarities.
- (1) Female pseudohermaphrodism linked to adrenal malfunction
- (2) Male pseudohermaphrodism, primarily a failure of the testes to function properly as masculinizing controllers of the differentiation of the gonaducts and external genitalia
- (3) Complete primary agonadism
- (4) Partial primary sterility
- (5) True hermaphrodism

The first and the second types are hereditary, showing characteristic patterns of familial distribution. The primary causes obviously are gene mutations. The occurrence of the last three is sporadic and their etiology (origin and cause) is right now an object of intensive and widespread search. The following paragraphs will be restricted to a consideration of the latter types, which show a close similarity of symptoms with the abnormalities of frogs derived from overripe eggs. The widely encountered deficiency in mental development of human cases is clearly related to incomplete embryonic development of the forehead, one of the most common abnormalities of the experimental frogs. If animal experimentation is applicable in the interpretation of pathologic development in man, then it may be concluded that gonadal agenesis combined with general teratology most likely results from physiologic damage (blastophthoria) to the egg or to early developmental stages [20, 21].

a) Complete gonadal agenesis. This type now usually designated as the *Turner syndrome*, is most frequently discovered in women who visit medical clinics because of primary amenorrhea (failure of menstruation to become established). Very often, though not always, other abnormalities are present: short stature, broad webbed neck, edema, and mental deficiency being most frequent. On exploration by laparotomy it is found that in typical cases gonads are absent or represented by vestigial peritoneal folds containing few germ cells or none. Tubes, uterus, and breasts—in short the prepuberal female secondary sex characters are present—although of *castrate type*. Interest in these cases was increased when about 1950 Barr and collaborators found that somatic cell nuclei of women contain a small nucleolus of so-called sex chromatin, which is not observed in corresponding male nuclei [22]. It was now realized that, while many of these "agonadal women" had the female sex chromatin, a majority did not. It was therefore concluded that some were genetic males. With appropriate hormone treatments the secondary sex characters of "pseudofemales" can be made to develop more fully. Even menstruation can be established. Many such patients later marry; they may become good housewives but, of course, they never bear children.

Even though the absence of gonads is usually discovered only at the time of puberty, the abnormality definitely is congenital. In a few cases the syndrome has also been obseved at birth. Edith Potter [23] reported the case of a girl born with webbed neck and edema of the lower extremities (Fig. 35). After death, at the age of 10 days, the heart was found severely abnormal; the gonads were sterile rudiments,

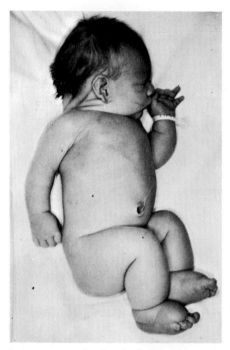

Fig. 35. Newborn apparently female baby with webbed neck and edema of the legs [23].

but the gonaducts, of normal female type, were differentiated into tubes, uterus, and vagina. Also, the external genitalia were typically female, but an investigation of some sections of adrenal tissue showed chromatin-negative nuclei, i.e. the male type (Fig. 36a). Why the gonaducts were of female type had, in the meanwhile, been solved by Jost [24] and Raynaud [25]. Early castration of rabbits and mice shows that, in mammals, female gonaducts differentiate in the absence of gonads as well as in the presence of ovaries (neutral type), while the male type differentiates only when induced by testicles. Therefore, the girl shown in Figure 35 acquired the female secondary sex characters simply because of absence of contrary induction.

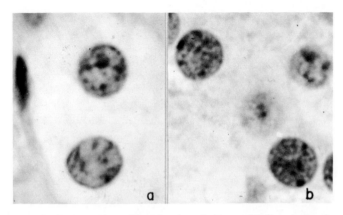

FIG. 36. Cross sections through adrenal tissues; (a) from baby shown in Figure 35: the nuclei are chromatin negative (male type); (b) from a normal female fetus of 14 cm length: the nuclei are chromatin positive (female type); both figures ×2000.

The enigmatic nature of the sex chromatin has led to many careful investigations and far-flung speculations. At the present, the most interesting interpretation is one proposed by Ohno, et al. [26], namely, that the chromatic nucleolus may represent a pyknotic segment of the paternal X chromosome. This may not yet be the final solution, although recent findings by Graham and Barr [27] on dimorphic sex chromatin in the opossum seem to fit well into this design.

b) Partial gonadal agenesis. A second type of sex deviation is seen in boys who, at early puberty, see the doctor because of a disturbing, feminine-type breast development. Beard and pubic hair are sparse. Often, somatic malformations and mental retardation are combined with the sexual abnormality. The testes are small and fail to give evidence of normal puberal enlargement. Biopsies show partially or completely sterile seminal tubules with resultant low fertility. The tubules contain mainly sustentacular cells of Sertoli. Occasionally, loci of complete spermatogenesis may be found, while in other cases, cortical remnants are seen [20]. This type of sex aberration is called the *Klinefelter syndrome.* In about half of the properly investigated cases the sex chromatin has been reported as of female type [20,21,28,29]. Studies in a male-infertility clinic indicate that 40 per cent of Klinefelter cases are of female nuclear sex, and that about 8.5 per cent of men suffering from subfertility are male and female type Klinefelters [30]. Ferguson-Smith, who investigated 663 male prepuberal children attending special schools for the mentally handicapped, found 1.2 per cent of chromatin-positive cases of "micro-orchidism" (female-type Klinefelter). Testicular biopsy showed mainly a drastic reduction of spermatogonia as compared with normal subjects of the same age [31]. This is clear evidence of the congenital nature of this aberration, and here again the analogy with the overripeness effect in the frog is quite obvious.

c) *Mixed types and true hermaphrodites.* If one assumes that postgenetic factors are responsible for gonadal agenesis, the cases of mixed male and female type can easily be understood. Evidently, agenesis varies in severity. Instead of maximal reduction, rudimentary testicles may develop with only a few immature seminal tubules and some clusters of interstitial cells (of Leydig). This causes *partial masculinization* either of the external genitalia or of the gonaducts. In cases of asymmetry—such as are often observed in experiments with frogs [32]—an oviducal tube will develop only on the side of the sterile gonad, while on the side of the rudimentary testicle, the oviduct is suppressed and the deferent duct instead, is maintained [28]. Some cases of *true hermaphrodism*, with fertile ovarian and testicular parts distributed in various patterns, possibly have a similar origin. One would assume usually that such individuals are genetically of female sex and that one gonad, or segments of one or both gonads, suffered a reduction in cortical inductive capacity that opened the way for testicular differentiation [20,29].

d) *Chromosomal aberrations.* In the last few years improved technics of investigation have revealed that the number of chromosomes in human beings with malformations of the degenerative types characterized above frequently deviates from the normal 46. A first clearly established case was that of mongolism, presented by a group of French investigators [33]. This syndrome varies widely in severity but always seems to involve underdevelopment of forehead and brain, and mental retardation. The presence of an extra chromosome of the type of the second smallest element (therefore called triplo 22 or triplo 21, according to different numbering systems) was established in over 30 cases. Even though this number is still relatively small, a fairly close correlation between syndrome and chromosomal aberration is evident.

In cases in which the abnormalities also concern the sex glands, the relationship is not so clear. The compilation in Table 1 is incomplete and also inadequate, because it cannot

TABLE 1. LIST OF CHROMOSOMAL TYPES WITH CORRELATED NORMAL AND ABNORMAL SEX CONDITIONS IN MAN*

Chromosomes		
Number	Pattern†	*Manifestations*‡
69	XXY	Male child
47	XXX	+"Super female" (1)
47	XXY	+Klinefelter (5)
46	XX	+*Normal female;* +*Turner* (1) +Klinefelter (1); + True hermaph. (3)
46	XY	−*Normal male;* −Female (1) −Turner (1); −Klinefelter (4) True hermaph. (1)
46–47	XXY-XX^m	+Klinefelter (1)
45–46	XX-XO^m	+Turner (3 or 4)
45	XO	−Turner (4); +Turner (1)
47	XX+(22)	+Female mongoloid (6)
47	XY+(22)	−Male mongoloid (8)
48	XXY+(22)	+Klinefelter mongoloid (2)

* Observations of Witschi (with Mikamo and Zellweger) and data from the literature [34–35].
† m, mosaic.
‡ Parentheses, number of cases; +, chrom. pos.; −, chrom. neg.

do justice to the individual pattern of all cases. In view of the fact that all genetically and chromosomally controlled cases of experimental sex reversal in vertebrates were found neither to be caused nor to be followed by chromosomal changes, it is not really surprising that most sex patterns can be realized under various normal and abnormal types of chromosome arrangements. Nevertheless, present data suggest some "preferred" combinations, such as XO with the Turner and XXY with the Klinefelter syndrome.

Characteristically, abnormal chromosome numbers are found in many types of pathologic cells such as cancer and leukemia and also in cells derived from overripe amphibian eggs [46]. This suggests that *cellular pathology may be the common cause for both chromosomal aberrations and the general teratologic nature of embryonic development*. Such an assumption does not disclaim the influence of abnormal chromosomal types on certain features of the developmental pattern, but, it offers a possible explanation for the origin of the chromosomal aberrations as well as for the lack of a close correlation between chromosomal types and patterns of sex deviation. It particularly helps to clarify the two characteristic features of the entire group: non-genetic origin and degenerative-type pathology. In the meanwhile, a study of overripe eggs of the rat has already given first results which suggest that non-disjunction of diads does occur and probably relatively frequently, in the second miotic division [11]. Obviously, present data are still insufficient to provide a clear and final causal explanation of the complexities of the many types of sex deviation in man, but interest in the study of teratologic development is now so widespread that decided progress in this field can be expected within the next few years.

REFERENCES

1. KNIEP, H. *Die Sexualität der Niederen Pflanzen.* Jena, Fischer (1928).
2. HARTMANN, M. *Die Sexualität.* Stuttgart, Fischer (1956).
3. LEDERBERG, J. *Harvey Lectures,* **52,** 69 (1959).

4. WITSCHI, E. *Development of Vertebrates.* Philadelphia, Saunders (1956).
5. WITSCHI, E. *Contrib. Embryol.* **32,** 67 (1948).
6. SHETTLES, L. B. *Am. J. Obstet. Gynecol.*, **66,** 235 (1953).
7. WITSCHI, E. *Science*, **130,** 372 (1959).
8. GOEBEL, K. *Biol. Zentr.*, **30,** 657 (1910).
9. BALTZER, F. *Verhandl. deut. zool. Ges.*, 273 (1928).
10. FURROW, C. L. *Zellf. mikr. Anat.*, **22,** 282 (1935).
11. WITSCHI, E. *Experientia*, **16,** 274 (1960).
12. BEDICHEK-PIPKIN, S. *Univ. Texas Publ.*, **5914,** 69 (1959).
13. DOBZHANSKY, T., and J. SCHULTZ. *J. Gene.*, **28,** 349 (1934).
14. SEILER, J. *Zellforsch.*, **15,** 249 (1921).
15. ANKEL, W. E. *Nat. & Mus.*, **6,** 273 (1929).
16. BENOIT, J. *Compt. rend. Acad. Sci.*, **177,** 1074 (1923).
17. CHANG, C. Y., and E. WITSCHI. *Proc. Soc. Exp. Biol. Med.*, **89,** 150 (1955); **93,** 140 (1956).
18. WITSCHI, E. *Cancer Research*, **12,** 763 (1952).
19. WITSCHI, E., and C. Y. CHANG. *Anat. Record*, **118,** 370 (1954).
20. WITSCHI, E. *Trans. Third Conf. Gest. Josiah Macy Found.*, C. A. Villee, Ed. 119 (1956).
21. WITSCHI, E., W. O. NELSON, and S. J. SEGAL. *J. Clin. Endocrinol. and Metabolism*, **17,** 737 (1957).
22. BARR, M. L. *Am. J. Human Genet.*, **12,** 118 (1960).
23. POTTER, E. *Pathology of the Fetus and the Newborn*, Chicago, Year Book Publ. (1957).
24. JOST, A. *Ann. Anat. Micr. et Morph. Exp.*, **39,** 395 (1950).
25. RAYNAUD, A. *Ann. Anat. Micr. et Morph. Exp.*, **39,** 336 (1950).
26. OHNO, S., W. D. KAPLAN, and R. KINOSITA. *Exp. Cell Research*, **18,** 415 (1959).
27. GRAHAM, M. A., and M. L. BARR. *Ann. Anat. Micr. et Morph. Exp.*, **48 bis,** 111 (1960).
28. GRUMBACH, M. M., and M. L. BARR. *Recent Progr. Hormone Research*, **14,** 255 (1958).
29. BUNGE, R. G. and J. T. BRADBURY. *J. Urol.*, **76,** 758 (1956).
30. FERGUSON-SMITH, M. A., B. LENNOX, W. S. MACK, and J. S. S. STEWART. *Lancet*, 167 (1957).
31. FERGUSON-SMITH, M. A. *Lancet*, 219 (1959).
32. WITSCHI, E. *A. mikr. Anat.*, **102,** 168 (1924).
33. LEJEUNE, J., R. TURPIN, and M. GAUTIER. *Ann. Gènètique*, **1,** 41 (1959).

34. Baikie, A. G., W. M. Court Brown, P. A. Jacobs, and J. S. Milne. *Lancet*, 425 (1959).
35. Böök, J. A., and B. Santesson. *Lancet*, 858 (1960).
36. Ford, C. E., P. A. Jacobs, and L. G. Lajtha. *Nature*, **181,** 1565 (1958).
37. Ford, C. E., K. W. Jones, O. J. Miller, U. Mittwoch, L. S. Penrose, M. Ridler, and A. Shapiro. *Lancet*, 709 (1959).
38. Ford, C. E., K. W. Jones, P. E. Polani, J. C. DeAlmeida, and J. H. Briggs. *Lancet*, 711 (1959).
39. Ford, C. E., P. E. Polani, J. H. Briggs, and P. M. F. Bishop. *Nature*, **183,** 1030 (1959).
40. Grumbach, M. M., A. Morishima, and E. H. Y. Chu. Soc. Pediatric Research, *30th Ann. Meet.*, 55 (1960).
41. Harnden, D. G., and J. S. S. Stewart. *Brit. Med. J.*, **2,** 1285 (1959).
42. Harnden, D. G., and C. N. Armstrong. *Brit. Med. J.*, **2,** 1287 (1959).
43. Hungerford, D. A., A. J. Donnelly, P. C. Nowell, and S. Beck. *Am. J. Human Genet.*, **11,** 215 (1959).
44. Lanman, J. T., and K. Hirschhorn. Soc. Pediatric Research, *30th Ann. Meet.*, 53 (1960).
45. Nilsson, I. M., S. Bergman, J. Reitalu, and J. Waldenström. *Lancet*, **2,** 264 (1959).
46. Beetschen, J., *Compt. rend. Acad. Sci.*, **245,** 2541 (1957).

AUSTRALIAN TREES AND HIGH BLOOD PRESSURE

By ROBERT C. ELDERFIELD
University of Michigan

"Enthusiasm arises from pride, hope, presumption, a warm imagination together with ignorance." DAVID HUME

THE SOCIETY of the Sigma Xi is dedicated to the promotion of scientific research, for the most part without regard to the prospect of immediate practical gains. As has been so often demonstrated in the past, such practical corollaries follow almost automatically once the basic ground work has been laid. Nowhere is this happy state of affairs more prominent than in the fundamental studies of the chemical constituents of the plant kingdom over the past hundred years. One need mention only the pioneering studies of Einhorn and particularly of Willstätter on the molecular architecture of cocaine, used for centuries by South American natives as a stimulant and later on as a local anesthetic. From these have resulted the host of synthetic modifications of the natural alkaloid which are now so familiar as local anesthetics and which have improved on nature to no inconsiderable extent. Yet I am sure that Willstätter, even in his fondest dreams, never envisioned such molecules as those of Novocain, Nupercaine and many others that could be cited. Rather, he was impelled by scientific curiosity as to what peculiar arrangement of the constituent atoms of the cocaine molecule was responsible for its characteristic action.

Since the end of the Second World War the study of the chemistry of natural products, particularly those of plant origin, has received new emphasis. This has been due to a number of stimuli. The spectacular success attending the development of the chemistry and physiology of cortisone has set in motion an intensive search for plant materials that can serve as precursors in the synthesis of this important substance. Undoubtedly the impact of reserpine as a hypotensive and tranquilizing agent, following from its traditional use in Indian medicine and folklore, has given impetus to the investigation of other naturally occurring substances. But, above all, one must credit the insatiable scientific curiosity of many workers in this field for the developments in structure determination, synthesis, and biogenesis which have been forthcoming in recent years.

Organic compounds arising from plant sources are conveniently classified according to their chemical nature. For example, the broad classes of substances known as terpenes, plant steroids, plant proteins, and alkaloids may be mentioned. Inasmuch as the alkaloids form the basis of the present discussion, it is not out of place to attempt a definition of such substances. No completely satisfactory, all-inclusive definition is possible; it will be sufficient to define an alkaloid as a nitrogenous substance usually of plant origin, usually possessing basic properties, usually optically active, and usually possessing some characteristic physiological action. Such a definition is not perfect, and exceptions to all of the above criteria can be cited.

The purpose of this discussion is twofold. In the first place, the investigations to be described furnish an excellent example of possible practical value which may ensue from studies undertaken with no immediate utilitarian objective in view, but are rather begun and continued merely for the satisfaction of scientific curiosity. Secondly, it is hoped that some insight into the methods and philosophy that the organic chemist brings to bear on the unraveling of molecular structures of complex natural products may be provided to those who are not particularly versed in this discipline.

At the time the work was undertaken, beyond a few scattered reports of possible therapeutic value of the substances under study, there was no practical incentive before us. This was even more true when these reputed therapeutic properties were disproved. Nevertheless, continuation of the investigation has furnished strong evidence for the hope that new and highly effective hypotensive agents may be at hand.

The Investigation of Plant Materials

In commencing an investigation of the constituents of a plant material certain well defined steps are indicated.

1. Procurement of adequate supplies of the plant material with particular emphasis on exact botanical identity. In this phase we are deeply indebted to our colleagues in the field and to the botanist and pharmacognoscist.

2. Development of suitable methods of extraction and separation of the plant material into its components in pure states. Two very powerful tools are available for this purpose: chromatographic adsorption on suitable adsorbants as originally developed by Tswett, and the more recent method of countercurrent distribution, which has been brought to its highest state of efficiency by Craig. Once the pure substances are at hand the problem of molecular structure demonstration can be attacked.

The classical attack on the problem of deducing the structure of a complex molecule, in general, involves breaking it down into smaller recognizable molecules by a variety of procedures. The fragments thus obtained are then pieced together, more or less in the fashion of a jig-saw puzzle, to provide a representation of the original molecule. To complete the structure demonstration, the substance under consideration is then synthesized from simpler substances by reactions, the course of which can be predicted accurately. Identity of the synthetic material with that of natural origin thus confirms the molecular architecture arrived at by degradation.

In recent years development of modern instrumental methods has greatly simplified the task of structure determination. Some feel that these advances have more or less taken the fun out of classical organic chemistry, but it cannot be denied that the advent of such tools as X-ray analysis and modern spectrographic methods has been largely responsible for the spectacular advances in the study of natural products.

Finally, a third dimension has been added to the structural expressions for these complex molecules. Whereas early investigators were, of necessity, content to make use of planar formulas, however inadequate these were recognized to be, the development of the technique of conformational analysis in recent years has provided means by which exact spatial locations of the constituent atoms in a given molecule may be assigned. It will not be possible to indulge in a rigorous treatment of this aspect of structural organic chemistry at this time.

In the material to be presented, examples of both the classical and the modern approaches will be found.

The Alkaloidal Constituents of Alstonia

Our investigations of the alkaloidal constituents of various species of Alstonia began in 1939. In casting about for a group of alkaloids that had received comparatively little attention, the alkaloid alstonine was found listed in an old

Merck Index. Subsequent investigations showed that it had been isolated by two or three earlier investigators from the tree bark of *Alstonia constricta* F. Muell., but that comparatively little was known concerning its chemistry. It also developed that extracts of the bark had enjoyed some reputation in the Chinese and Philippine literature as antifebretic agents in the treatment of malaria. Indeed, extracts of the bark were at one time listed in the British Pharmacopeia. The next step was to secure a supply of the bark and, in the course of this, another interesting item emerged. One of the major suppliers of crude drugs in this country reported that it had a few hundred pounds of *A. constricta* bark in the warehouse of a company recently acquired in Cincinnati. It also appeared that a tincture of the bark had at one time been used as a component of a cough syrup.

Accordingly, an investigation of alstonine, the major alkaloidal constituent of *A. constricta* bark, was undertaken in association with Nelson J. Leonard, now of the University of Illinois. The myth of the effectiveness of both alstonine and an extract, consisting of the total alkaloids of the bark as antimalarials, was quickly disposed of through the cooperation of Dr. John Bauer of the Rockefeller Foundation. Alstonine showed a quinine equivalent somewhat less than one-half against avian malaria, and the total alkaloid fraction was considerably less active. Nevertheless, the chemical investigation was continued.

Alstonine, $C_{21}H_{20}N_2O_3$, was first investigated in detail by Sharp [1-3] who noted the presence of one methoxyl, OCH_3, group as a methyl ester, one basic and one chemically inert nitrogen, and one chemically inert oxygen.

Sharp also commenced the process of breaking down the alstonine molecule into simpler pieces. Oxidation of the alkaloid with permanganate resulted in the formation of oxalylanthranilic acid, a degradation product the formation of which is considered indicative of the presence of an indole ring system. Sharp also applied the procedure of selenium dehydrogenation, which had been utilized with striking results by Ruzicka in unraveling the structure of the large group of substances known as the terpenes, to alstonine. In this procedure the material under examination is heated with selenium, during which hydrogen is removed, sometimes with concurrent rupture of chemical bonds; and a more easily identifiable completely aromatic substance, which reflects the skeleton of the parent compound, is obtained. The product of the dehydrogenation of alstonine was a substance called by Sharp *alstyrine*. Although he was unable to assign a structure to alstyrine at the time, subsequently alstyrine was to supply the key to the structure of alstonine, just as Ruzicka had been able to assign structures to many of the terpenes on the basis of identification of the products of their dehydrogenation. The status of the knowledge of the structure of alstonine at the time our investigations began may then be represented as follows.

Tetrahydro-alstonine $C_{21}H_{24}N_2O_3$ $\xleftarrow{\text{Pt-H}_2\ +2H_2}$ Alstonine $C_{21}H_{20}N_2O_3$ $\xrightarrow[\text{hydrolysis of ester}]{\text{NaOH-H}_2O}$ Alstoninic Acid $C_{20}H_{18}N_2O_3$

Se ↙   KMnO$_4$ ↘

Alstyrine $C_{19}H_{22}N_2$

Oxalylanthranilic Acid (COOH, NHCOCOOH on benzene ring)

Further information as to the skeleton of alstonine was provided by Leonard and Elderfield [4], who definitely demonstrated the presence of a β-carboline nucleus in alstonine by application of the classical methods of distillation with zinc dust and fusion with alkali, procedures by which a large molecule may frequently be degraded to identifiable fragments. In this instance the products were harman and norharman (β-carboline), both of which had been adequately characterized previously by Kermack, Perkin, and Robinson during the course of their study of the alkaloid, harmine. Another diagnostic reaction, which has also had fruitful application in the terpene field, involves reduction with sodium and an alcohol. Esters are reduced to primary alcohols, and isolated carbon-to-carbon double bonds are unaffected by this reagent. However, if the carbon-to-carbon double bond is in conjugation with an ester function, it is also reduced with formation of a saturated alcohol:

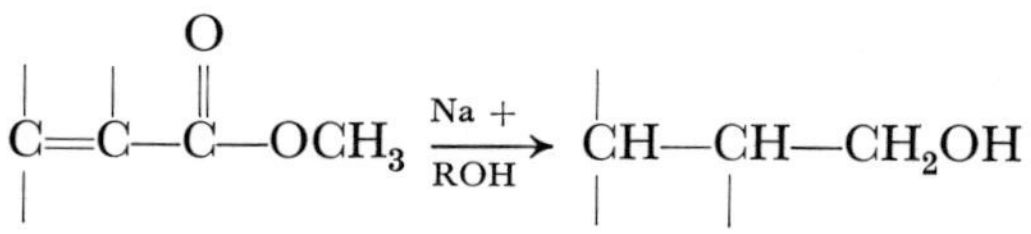

$$\underset{|}{C}=\underset{|}{C}-\overset{O}{\underset{\|}{C}}-OCH_3 \xrightarrow[ROH]{Na+} \underset{|}{C}H-\underset{|}{C}H-CH_2OH$$

By this reaction the presence of such an unsaturated ester in alstonine was demonstrated. Assignment of a partial structure to alstonine may then be made as indicated.

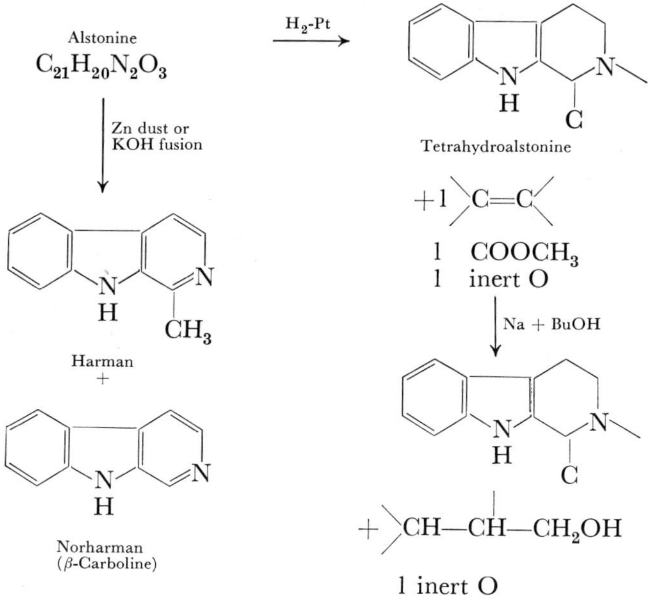

Alstonine
$C_{21}H_{20}N_2O_3$

$\xrightarrow{H_2\text{-Pt}}$ Tetrahydroalstonine

↓ Zn dust or KOH fusion

Harman
+
Norharman
(β-Carboline)

+1 C=C
1 COOCH$_3$
1 inert O

↓ Na + BuOH

+ CH—CH—CH$_2$OH

1 inert O

At this time further work was postponed because of the more pressing demands of World War II. The investigation was resumed after the cessation of hostilities, and now Australia enters the picture. The original supply of bark having been consumed, arrangements were made with Dr. J. R. Price of the Australian Commonwealth Scientific and Industrial Research Organization for the collection of additional bark from northern Australia to which it is indigenous. It is a pleasure to acknowledge the hearty cooperation of Dr. Price, both in this instance and in providing other species of Alstonia to be discussed in the sequel.

In the interim other advances had been made, both from the chemical and the physiological standpoint. The potent properties of an extract of *Rauwolfia serpentina* Benth as a blood pressure-lowering and tranquilizing agent had been known to the people of India for years, and an investigation of this plant had been undertaken in the laboratories of the Ciba Company and elsewhere [5]. The major active principle of Rauwolfia, reserpine, was isolated and, in addition, a second alkaloid, serpentine, which appeared to be isomeric with alstonine, was found, among others. Further, reserpine had been located in the root bark of *Alstonia constricta*, although we have obtained no evidence of its occurrence in the tree bark. Finally, tetrahydroalstonine had been found in several varieties of Rauwolfia. It thus became apparent that hypotensive agents might, in all probability, be present in the Alstonia species.

Advances had also been made on the problem of the structural chemistry of alstonine and its derivatives. Karrer and Enslin in Switzerland [6] had demonstrated the structure of alstyrine which has subsequently been confirmed by synthesis [7]. Woodward and his associates [8] had elucidated the structure of sempervirine, an alkaloid from *Gelsemium sempervirens* and, in particular, had recorded its spectrographic properties. At this time it will be sufficient to say that the rather unusual bipolar structure resonating with a nonpolar structure, which is present in sempervirine, is a specific chromophoric group as shown in the ultraviolet portion of the spectrum. This highly characteristic chromophore is shown schematically in the following structures.

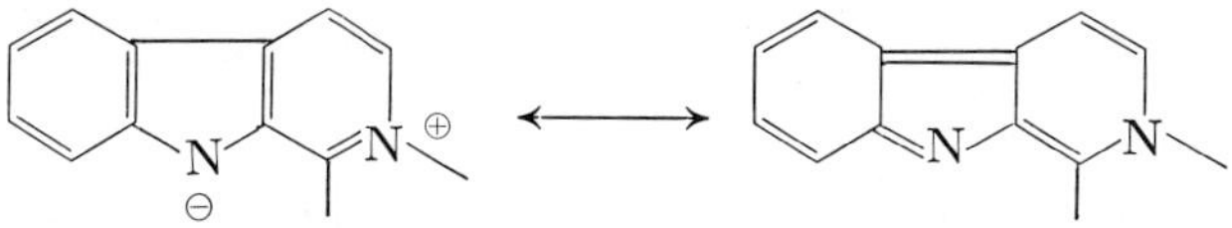

Indeed, the contribution of the nonpolar form with its quinonoid arrangement of double bonds results in absorption in the visible portion of the spectrum with resultant development of color in the substances. Both alstonine and its stereo-isomer, serpentine, are colored, and their absorption spectra show close resemblances to that of sempervirine. It therefore appeared to be a reasonable deduction that the above chromophore was present in the alkaloids under consideration [9]. These considerations, taken in conjunction with the degradation studies previously discussed, led to assignment of the following structure for serpentine [10,11].

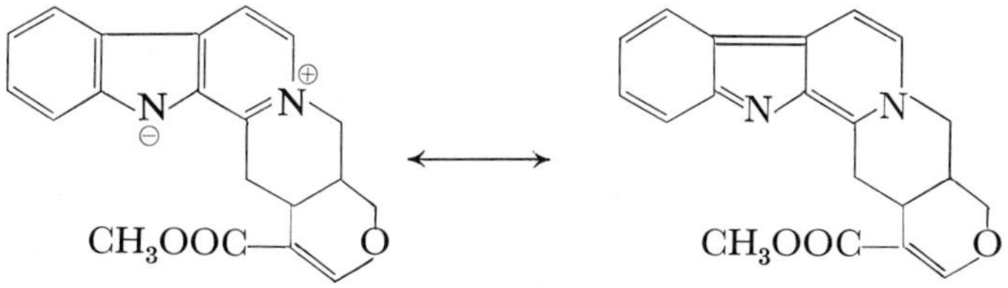

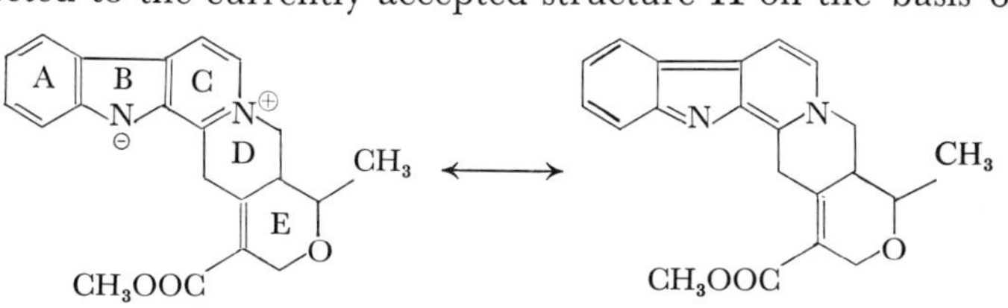

However, alstonine contains a C-methyl group that cannot be accommodated by the above structure. Consequently, structure I was advanced for alstonine by Elderfield and Gray [12]. Subsequently, both these structures have been corrected to the currently accepted structure II on the basis of spectral evidence [13], particularly in the infrared, which provided a basis for assignment of the arrangement in the oxygen heterocyclic ring E. The various reactions and degradations of alstonine then become easily understood and the pieces of the jig-saw puzzle fit logically into place.

At this point it may be profitable to discuss briefly the relationship of alstonine to its stereoisomer, serpentine. At the risk of oversimplification it can be stated that current stereochemical theory permits the existence of two arrangements in space of the constituent atoms present when two saturated carbocyclic rings are fused. This is most conveniently illustrated in the simple case of the decahydronaphthalenes.

III IV

The hydrogen atoms at the bridgeheads may lie in one or the other of two conformations: they may bear a *cis* relationship to each other as shown in III or they may bear a *trans* relationship as in IV. Assignment of specific arrangements in such instances rests on appropriate degradation reactions whereby simpler compounds of known configuration are obtained. In general, such differences in conformation manifest themselves in differences in chemical reactivity between the members of the isomeric pairs and in differences in their infrared absorption spectra. Arguments such as these have been applied to the alstonine-serpentine isomerism question [14]. One specific reaction enables a differentiation to be made between isomers carrying *cis*-hydrogens at the D-E ring juncture (positions 15 and 20) of compounds of this general type and those carrying *trans*-hydrogens. This involves a comparison of the rates of catalytic dehydrogenation of the tetrahydro derivatives to the parent alkaloids in the presence of palladium and a hydrogen acceptor such as maleic acid. Presumably, the difference in rates can be ascribed to the fact that the conformation of the more reactive isomer permits a better "fit" on the surface of the catalyst with resultant faster reaction. This isomerism also manifests itself in differences in physiological response shown by the two isomers. The interrelationship of alstonine and serpentine then may be represented as follows.

[Note added in proof: The stereochemical assignments to the D/E ring juncture in alstonine and serpentine as given in the original lecture were based on information available at that time [14]. Very recently evidence has been presented that this earlier assignment is in error and must be reversed. Thus the D/E ring fusion in alstonine is now *cis* and that in serpentine is *trans*. The accompanying formulas have been corrected accordingly to take this into account. See E. Wenkert, B. Wickberg, and C. L. Leicht, *J. Am. Chem. Soc.*, **83,** 5037 (1961), and M. Shamma and J. B. Moss, *J. Am. Chem. Soc.*, **83,** 5038 (1961). This reversal of assignment should be borne in mind in the discussion of the relation between stereochemical configuration and physiological action of the substances as appears below.]

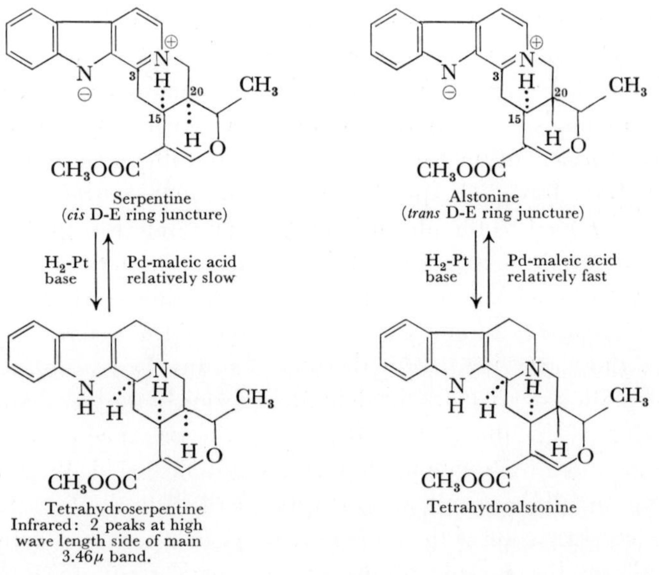

Serpentine
(*cis* D-E ring juncture)

Alstonine
(*trans* D-E ring juncture)

H$_2$-Pt base | Pd-maleic acid relatively slow

H$_2$-Pt base | Pd-maleic acid relatively fast

Tetrahydroserpentine
Infrared: 2 peaks at high wave length side of main 3.46μ band.

Tetrahydroalstonine

Pharmacological data of a limited nature concerning these substances are at hand. Alstonine has been investigated by Wakim and Chen [15] who report that the hydrochloride of the alkaloid lowers the blood pressure of anesthetized dogs, cats, and rats. It also shows an adrenolytic effect in that the response of adrenaline in raising the blood pressure is reduced. Similar findings have also been noted by Page [16]. Serpentine has been studied by a number of investigators and displays weaker hypotensive properties. Serpentinine, which may be a third isomer of alstonine, has also been shown to exert a hypotensive effect in animals.

In any event, there appeared to be good reason to believe that further investigation of other Alstonia species might well provide other and more useful hypotensive agents. The species comprises a rather large group belonging to the family of Apocynaceae and has been the subject of comparatively little chemical attention. The alkaloids that have been reasonably well characterized are listed in Table 1.

TABLE 1. ALKALOIDS OF OTHER ALSTONIA SPECIES

Species	Alkaloids Isolated	References
A. constricta F. Muell. (tree bark)	Alstonine, alstoniline, alstonidine	17–20
A. constricta F. Muell. (root)	Alstonine, alstonidine, reserpine, α-yohimbine	21–23
A. macrophylla Wall	Villalstonine, macralstonine, macralstonidine	24
A. villosa Blum	Villalstonine	24
A. somersetensis F. M.	Villalstonine, macralstonidine	24
A. verticulosa F. Muell.	Echitamine	24
A. scholaris R. Br.	Echitamine, echitamidine	25
A. angustiloba Miq.	Echitamine	25
A. congensis Eng.	Echitamine, echitamidine	24, 26
A. gilletii De Wild	Echitamine	25
A. spathulata Blume	Echitamine	25

In extending our investigation, we naturally turned to the remaining alkaloids of *A. constricta*, since a supply of this species was at hand. The chemistry of alstoniline, a physiologically inactive alkaloid, has been elucidated and a synthetic route to it and its derivatives has been opened up by which it is hoped to produce substances of pharmacological interest.

The structure of alstoniline was arrived at by the following transformations [27].

Synthesis of alstonilinol which confirms the structure assigned to alstoniline has been accomplished as follows [28]. Obvious variations of these synthetic procedures may well produce substances of pharmacological and therapeutic interest.

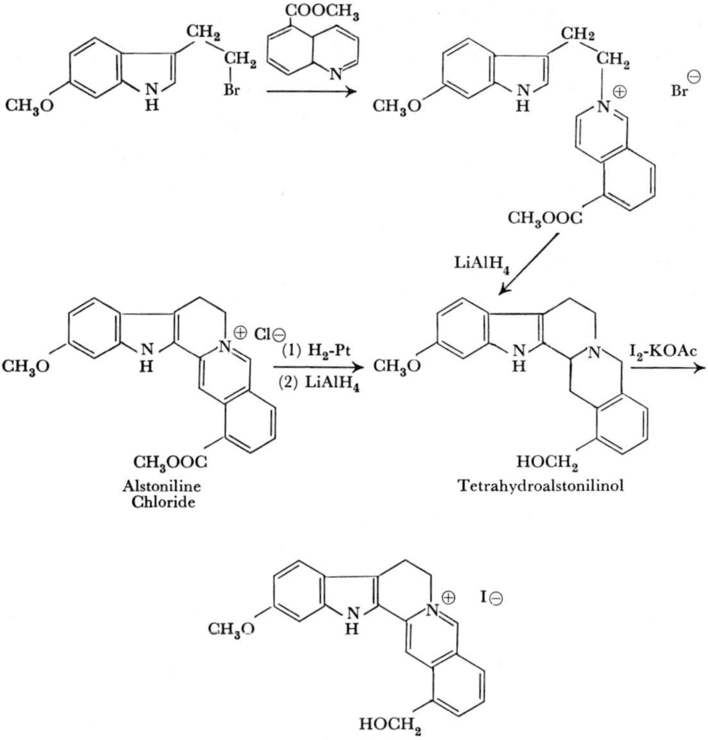

Alstoniline Chloride

Tetrahydroalstonilinol

Alstonilinol Iodide
Identical with substance prepared from
natural alstoniline

The alkaloid, alstonidine, is interesting on two counts. From the biogenetic point of view, it represents a new type of indole alkaloid in which ring D of the parent pentacyclic structure is open. From the chemical point of view it represents an instance of elucidation of structure almost entirely by physical methods without recourse to classical organic degradative methods [29].

Titration of alstonidine in 66 per cent dimethylformamide indicates a molecular weight of 382.5 and a pK_a' of 5.95 compared with that for harman of 6.10. Molecular weight by X-ray unit cell dimensions and density is 380.1. Molecular weight calculated for $C_{22}H_{24}N_2O_4$, to which the elementary analytical data correspond, is 380.4. Functional group analysis indicates the presence of one N-methyl, one C-methyl, and one O-methyl group. The O-methyl group is characterized as that of a methyl ester by alkaline hydrolysis to alstonidinic acid from which alstonidine can be obtained on methylation with diazomethane.

The ultraviolet spectra of alstonidine, of harman and of indole-N-methylharman are shown in Figure 37. From these

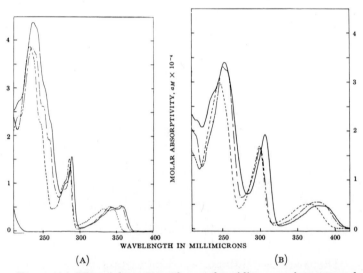

FIG. 37. (a) Ultraviolet spectra of ——— alstonidine, – – – harman, and — - — indole-N-methyl harman in methanol at 0.05 M potassium hydroxide [29]. (b) Same as (a) except in methanol at 0.05 M hydrochloric acid. Ultraviolet spectra were obtained on a Cary Model 14 spectrophotometer [29].

it can be concluded that the indole nitrogen of alstonidine is methylated and that an oxygenated substituent is absent in the harman ring system of the alkaloid.

The presence of a methyl ester conjugated with a vinyl ether is indicated from the ultraviolet absorption at 235 mμ and from absorption bands at 5.89 and 6.14 μ in the infrared

(Fig. 38). Within experimental error these maxima are identical with those shown by tetrahydroalstonine and indicate a similar arrangement in both molecules insofar as ring E is concerned.

Further, the infrared spectrum of alstonidine shows the presence of a strongly hydrogen-bonded hydroxyl group, the molar absorptivity of which is independent of the concentration. This absorption occurs at 3.18 μ. Likewise, the expected free hydroxyl band at 2.79 μ is very weak. On the other hand

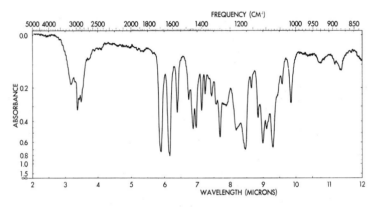

FIG. 38. Infrared spectrum of alstonidine in chloroform; 0.145 M in 0.11 mm path; Perkin-Elmer Model 21 spectrophotometer [29].

the infrared spectrum of alstonidine hydrochloride shows similar hydrogen bonding, except that the band is at 3.00 μ. These observations are interpreted as indicating that in alstonidine the hydroxyl group is involved in a strong hydrogen bond with the basic nitrogen. On the contrary, in the hydrochloride in which the basic nitrogen is protonated, a weaker hydrogen bond occurs between the hydroxyl group and the ether oxygen. Although hydroxyl absorption is very weak, the presence of such a group can be demonstrated by preparation of an acetyl derivative of alstonidine. This interpretation is summarized in Chart 1.

Chart 1

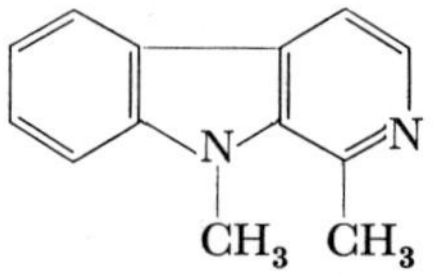

Indole-N-methylharman
$C_{12}H_9N_2$
(On basis of pK_a and UV absorption)

which leaves a $C_{10}H_{15}O_4$ fragment to be accounted for

$C_{10}H_{15}O_4$ can be expanded to on the basis of UV and IR data

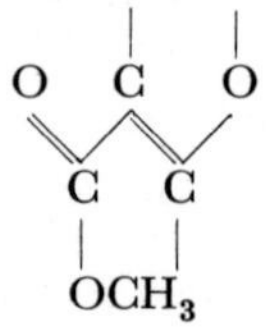

This leaves $C_6H_{12}O$ to be accounted for

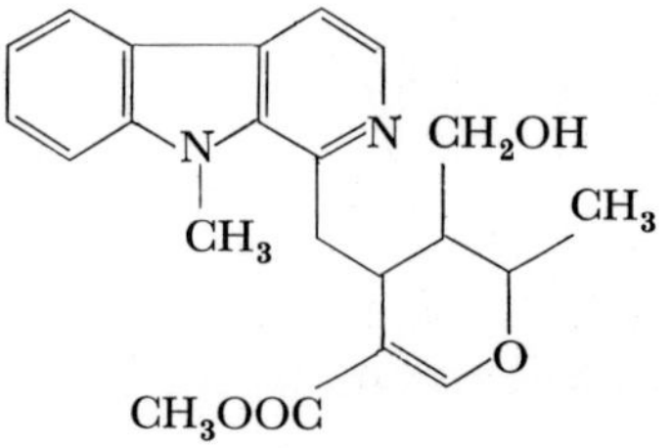

Alstonidine

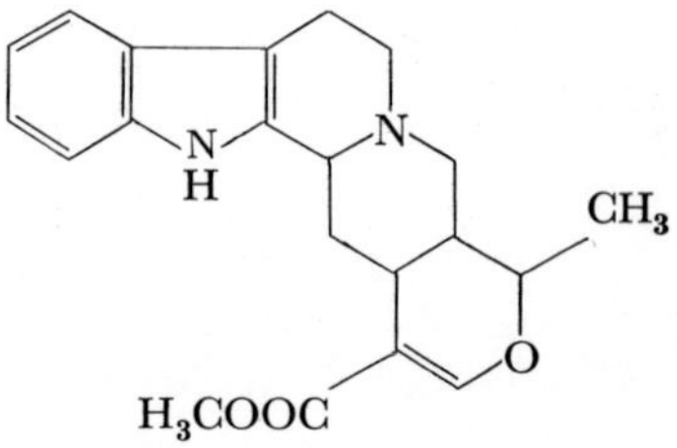

Tetrahydroalstonine

Finally, again through the cooperation of Dr. Price, a supply of the bark of *Alstonia muelleriana* has been obtained. From this, four new alkaloids at present unnamed, have been isolated [30] and application of paper chromatography has provided evidence of the presence of some twenty more alkaloids in the total alkaloid fraction of the bark. Separation and isolation of the four currently recognized alkaloids was accomplished by a rather long and tedious process involving both column elution chromatography and the more recent countercurrent distribution procedure of Craig [31]. The principles and operation of elution chromatography are too well known to warrant further discussion. In the present

CHART 2. EXTRACTION OF *Alstonia Muelleriana*

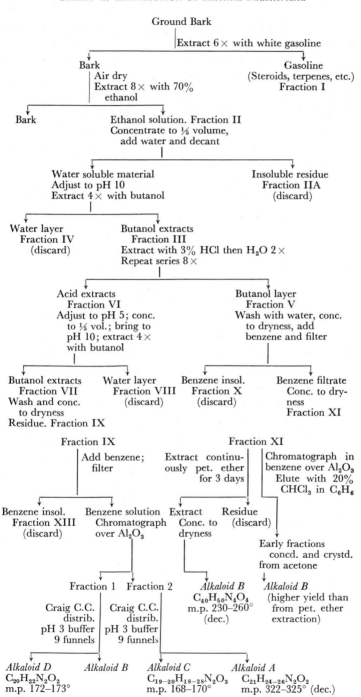

instance, use of the technique provided a rough separation of the alkaloids.

The countercurrent method may not be so familiar. In principle it applies the basic theories of fractional distillation to the separation of solid materials by taking advantage of their basic or acidic properties. By proper choice of a buffer system a mixture of substances, provided they are not exactly neutral, may be preferentially distributed between an aqueous buffer phase and a nonaqueous inert solvent. If this distribution is repeated as many times as may be required in a given instance, the net result is the application of a number of theoretical plates, as in a fractionating column, with resultant separation of the components of the mixture. The mathematical treatment of the procedure involves simple binomial theory exactly as with a column for the separation of liquids. In the present case nine theoretical plates were applied to the separation, but it is obvious from the qualitative results of paper chromatography that a more extensive system will be required for isolation of the remaining constituents. The isolation scheme heretofore used is illustrated in Chart 2.

Information on the chemistry of the four alkaloids is scanty at present. However, preliminary pharmacological data are of some interest and provide the basis for the hope that superior hypotensive agents may be forthcoming. At a dose level of 1 mg per kg, a very crude, highly impure total alkaloid fraction of *A. muelleriana* produced a gradual fall of mean arterial blood pressure reaching a maximum of 32 per cent in one hour in an anesthetized dog. There were no significant changes in respiration, intestinal motility, or the electrocardiogram [32]. This appears to be highly significant. For comparison, the result of a similar injection of 1 mg of *pure* reserpine is shown in Figure 39. It seems that some component of the crude total alkaloid fraction is much more potent than pure reserpine and also faster acting. To which alkaloid this action is due remains to be seen. Alkaloids B,

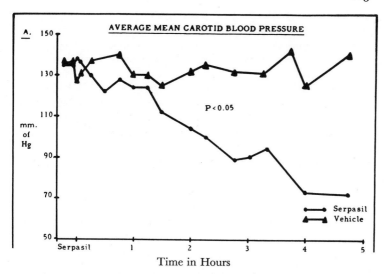

Fig. 39. The effect of reserpine (Serpasil) (1 mg per kg) on average mean carotid blood pressure in dogs (Trapold and co-workers [33]).

C, and D are not responsible for this strong hypotensive effect, and the available amounts of the remaining alkaloids have been insufficient for pharmacological examination.

REFERENCES

1. SHARP, T. M. *J. Chem. Soc.*, 287 (1934).
2. SHARP, T. M. *Ibid.*, 1227 (1934).
3. SHARP, T. M. *Ibid.*, 1353 (1938).
4. LEONARD, N. J., and R. C. ELDERFIELD. *J. Org. Chem.*, **7**, 556 (1942).
5. For a detailed treatment of the Rauwolfia alkaloids, see WOODSON, YOUNGKEN, SCHLITTLER, and SCHNEIDER. *Rauwolfia: Botany, Pharmacognosy, Chemistry, and Pharmacology*. Boston, Little, Brown and Co. (1957).
6. KARRER, P., and P. ENSLIN. *Helv. Chim. Acta*, **32**, 1390 (1949); **33**, 100 (1950).
7. LEE, T. B., and G. A. SWAN. *J. Chem. Soc.*, 771 (1956).
8. WOODWARD, R. B., and B. WITKOP. *J. Am. Chem. Soc.*, **71**, 379 (1949); R. B. WOODWARD, and W. M. MCLAMORE. *Ibid.*, **71**, 379 (1949).

9. SCHWARZ, H. *Experientia*, **6,** 330 (1950).
10. SCHLITTLER, E., and H. SCHWARZ. *Helv. Chim. Acta*, **33,** 1463 (1950).
11. At this time the presence of the double bond conjugated with the carbomethoxyl group in serpentine was not recognized.
12. ELDERFIELD, R. C., and A. P. GRAY. *J. Org. Chem.*, **16,** 506 (1951).
13. JANOT, M.-M., and R. GOUTAREL. *Bull. Soc. Chim., France*, 588 (1951); E. SCHLITTLER, H. SCHWARZ, and F. BADER. *Helv. Chim. Acta*, **35,** 271 (1952); F. BADER. *Ibid.*, **36,** 215 (1953).
14. WENKERT, E., and D. K. ROYCHAUDHURI. *J. Am. Chem. Soc.*, **78,** 6417 (1956); **79,** 1519 (1957).
15. WAKIM, K. G., and K. K. CHEN. *J. Pharmacol. Exp. Therap.*, **90,** 57 (1947).
16. Private communication from Dr. I. H. PAGE.
17. SHARP, T. M. *J. Chem. Soc.*, 287 (1934).
18. SHARP, T. M. *Ibid.*, 1353 (1938).
19. ELDERFIELD, R. C., and W. L. HAWKINS. *J. Org. Chem.*, **7,** 573 (1942).
20. ELDERFIELD, R. C., and E. SCHENKER. Unpublished work.
21. CROW, W. D., and Y. M. GREET. *Australian J. Chem.*, **8,** 461 (1955).
22. CURTIS, R. G., G. J. HANDLEY, and T. C. SOMERS. *Chemistry & Industry*, 1598 (1955).
23. Private communication from G. W. SVOBODA.
24. SHARP, T. M. *J. Chem. Soc.*, 1227 (1934).
25. GOODSON, J. A. *Ibid.*, 2626 (1932).
26. MONSEUR, X., and M. L. VANBEVER. *J. pharm. Belg.*, 93 (1955).
27. ELDERFIELD, R. C., and S. L. WYTHE. *J. Org. Chem.*, **19,** 683 (1954); R. C. ELDERFIELD, and O. L. MCCURDY. *Ibid.*, **21,** 295 (1956).
28. ELDERFIELD, R. C., and B. A. FISCHER. *Ibid.*, **23,** 949 (1958).
29. BOAZ, H., R. C. ELDERFIELD, and E. SCHENKER. *J. Am. Pharm. Assoc., Sci. Ed.*, **46,** 510 (1957).
30. ELDERFIELD, R. C., and R. E. GILMAN. Unpublished work.
31. CRAIG, L. C., and D. CRAIG. *Technique of Organic Chemistry*, A. Weissberger, Ed., vol. 3, p. 259. New York, Interscience Publishers, (1950).
32. Private communication from Dr. K. K. CHEN.
33. TRAPOLD, J. H., A. J. PLUMMER, and F. F. YONKMAN. *J. Pharmacol. Exp. Therap.*, **110,** 205 (1954).

MORPHOGENESIS IN PLANTS—
A NEW APPROACH

By Ralph H. Wetmore
Harvard University

The phenomenon of vegetative reproduction is commonplace knowledge to most people. We know that fruit trees can be propagated by grafting a small branch or a bud from a desired tree onto a less favored tree, usually of the same or a closely related species. By this means the grafted scion, whether small branch or bud, will grow and bear only the desired fruit, no matter how old or how large the scion may eventually become. The perpetuation of particular varieties of strawberries is assured by planting pieces of runners with a joint or node, of potatoes by utilizing certain "eyes" or buds borne on the tuber. The standard method of getting new fields of sugar cane started, or new stands of bananas, is to plant segments of the stems, segments which contain only a node or two. The housewife who wishes to multiply house plants, whether African violets or pelargoniums, knows that she can do so by taking a leaf of a fleshy plant or a section of a leafy stem, rooting these by appropriate tactics and then planting each newly rooted portion as the beginning of a new plant. These and like methods represent ways of obtaining new or preferred plants from old ones without resorting to the variable progeny one learns to expect from seeds.

In these methods of vegetative propagation, the size of the "seed" piece and also its place of origin raise only the question of how best technically to perpetuate it. For example, as Ball has already demonstrated for certain flowering plants [1], we find that a very small piece from the apex of a shoot (0.25 to 0.5 mm) will grow, if planted on an adequate nutrient medium. If the segment forms even a single root, a whole plant is assured. By contrast, since apices of ferns root easily, whole plants are readily obtained from apical pieces no longer than 0.25 mm in length, and even as short vertically as one cell thick. But if a piece is taken from the stem elsewhere than from the shoot apex, though grown in sterile nutrient medium, it does not ordinarily produce organized growth. True, it grows, but the aggregation of cells is usually of an unorganized, more or less homogeneous parenchyma, commonly spoken of as callus [2–6].* On an adequate medium, cell proliferation followed by cell enlargement seems to occur throughout the mass without the cellular differentiation into the tissue types found in the growing plant. If we should become able, at will, to introduce the fundamental patterns of organization into isolated plant calluses, we should then have a working knowledge of plant morphogenesis. Recently some progress has been made in bringing about centers of organization or buds in callus masses. Miller and Skoog [7,8], by the judicious use of an auxin (indole-3-acetic acid), and kinetin (6-furfuryl-aminopurine, a compound derived from deoxyribonucleic acid or DNA, the natural substance forming the background of chromatin) [9–11], have induced bud formation in callus

* A recent comprehensive work is now available on methods for and significance of sterile nutrient tissue, organ, and cell cultures for plants: Gautheret, R. J., *La Culture des Tissus Végétaux*. Paris, Masson et Cie. (1959).

derived from tobacco pith. These buds, which happen to produce both leaves and roots, make even callus a tissue that can be turned to vegetative propagation in tobacco. In time, this method may be the starting place for similar practices with other species.

The Limits and Limiting Factors in Early Morphogenesis

If we now return to the problem of how small a piece one can use in vegetative propagation, we are forced to consider the potentialities of single cells. In other words, though we are aware that vegetative reproduction is so common as to constitute a major method, by one device or another, for propagation of plants of the desired type, it has not been clear until now how small a piece can be taken to obtain an entire plant—whether one cell is enough or not.

Attention can be directed to some work from Prof. F. C. Steward's laboratory at Cornell University [12-14]. Cultures were started from uniformly-sized tissue samples taken from the specific region of the secondary phloem of tap roots of carrots. If left in the root the cells of these samples would normally have shown no subsequent growth or other changes. When the samples were grown in specially prepared glass culture flasks supplied with a basic liquid medium supplemented by 10 per cent coconut milk and the whole steadily but slowly rotated on a wheel, it was noticed that the transplanted disc grew into a callus and also that the surrounding fluid medium became cloudy. An examination of this fluid showed that, around the growing explants, were great numbers of living carrot cells, some single and some in small clusters. These cells were obviously liberated from the surface of the growing explants. Interesting results were noted in the growth of these cell cultures. The initial single cells or few-celled clusters achieved the irregular masses of the multiple cell stage by different tactics. For example, many cells repeatedly divided into equal halves, giving aggregations of isodiametric cells. Other cells, however, enlarged to many times their original size, reaching dimensions of 300 to 400 micra, but the contents of these large cells usually showed numerous nuclei, indicating that several nuclear divisions had occurred. The protoplasm of the large cell, however, eventually divided, one nucleus to each cell, so that a group of small cells could be recognized within the original cell wall. In other cases, single free cells elongated, each producing a filament which then divided transversely. Still other cells, especially some of those derived from the cotyledons of peanut, budded like yeast cells.

By whichever method individual cells multiplied—whether by equational division, by internal nuclear division in much enlarged cells which only eventually became multicellular, by filamentous growth which later divided into cells, or by budding—they all eventually achieved a multicellular, irregular shape. The subsequent growth of these cell aggregates was quite uniform and followed a general pattern. In each irregular three-dimensional cell mass a central aggregation of cells became early distinguishable, in that the component cells underwent those cytological changes recognized as steps in the differentiation of xylem tracheids or vessel elements. These central cells were surrounded by others which may well prove to be phloem-conducting units or sieve tube elements. The latter are, however, more difficult to recognize cytochemically or histochemically. Just peripheral to the xylem part of the nodule of vascular elements, orderly divisions occurred, preponderantly in a plane tangential to the outside of the whole cell aggregate, thereby giving a cambial-like zone. This more or less spherical-shaped, central nodule of cells appears characteristic of all these aggregates. It has its incidence only in the central region of the mass when, and if, a certain size is reached, as though the stimulus that induces the formation of these vascular tissues is somehow related to a particular internal environment that has been achieved for these centrally located cells (and only possible then). Is it possible that a lowered oxygen tension or a lowered level of some or all nutrients is important in the selection of such a central group? Attention must be given to this vascular center since to it all subsequently initiated appendages, root or shoot, are connected.

While initiation of a root may occur on a rotating wheel, a shoot never seems to develop there. When the cultures in this seemingly terminal stage of development on the wheel were transferred to stationary tubes and the liquid medium was made solid by agar, a root, already initiated on the wheel, soon continued growth or a new one was initiated. The root was followed by the beginnings of a shoot axis, at about 180° away from the root. This shoot axis also proved to have direct vascular connection with the nodular mass and thence to the root system. On the shoot axis, leaf primordia appeared in orderly fashion. The embryonic plant, with a root and shoot, continued its growth and development until it was a carrot plant with characteristically dissected leaves and tap root.*

Similar findings occurred with callus from potato tubers and from peanut cotyledons; multicellular aggregates originated in the same diverse ways as from carrot cells. As yet, however, whole plants of potato and peanut have not been achieved. This seems to be a question of time only, for the orderly developmental changes and procedures, in so far as they have gone, duplicate those of the carrot.

It is a remarkable fact that the steps followed by the single cell in becoming an embryonic plant resemble essentially those presented by the developing fertilized egg of any of the vascular plants. For example, in the angiosperms, the first division of the fertilized egg separates a suspensor cell from

* Such plants have recently been brought to flowering: MITRA, MAPES, and STEWARD, *Am. J. Botany*, **47**, 363 (1960).

its twin, the proembryonal cell. From the latter the embryo proper takes form by equational cell multiplication. From the suspensor cell, an elongating filamentous structure arises, which serves, in its elongation, to push the terminal, spherical embryo down into the special nutritive tissue, the endosperm. This endosperm develops from the unique pattern of fertilization in the angiosperms, in which the second male nucleus unites with polar nuclei to form the endosperm nucleus. Characteristically, the endosperm nucleus divides before the fertilized egg. Moreover this division, and also subsequent ones, are ordinarily not accompanied by cell-wall formation for some time, so that the young endosperm tissue in the embryo sac appears to have numerous free nuclei even as do many enlarged cells of the carrot callus. This multinucleate protoplasm of the embryo sac surrounds the multicellular, spherical embryo at the end of the filamentous suspensor. In consequence we see the specialized, nutritive endosperm growing and developing at the expense of the enveloping ovular tissue and gradually taking its place. The contained embryo is nourished by the surrounding endosperm, part of which may remain in the free nuclear or "milk" stage for some time. Gradually, wall formation or cytokinesis takes place around each free endosperm nucleus; thus, the walled-off solid endosperm forms at the expense of the "milk" or free nuclear endosperm. It seems, therefore, that the ovular tissues are digested and elaborated as endosperm which, in turn, becomes the nutritional background for the enclosed embryo. Clearly the formation of endosperm must involve biochemical changes, not just an increase in the number of cells involved in the nutrition of the embryo.

Steward and his co-workers* have ascertained that those endosperms so far investigated do contain growth-promoting

* See [15] for an excellent review of the nutritional significance of coconut milk and other endosperms and of the use of coconut milk in studies of morphogenesis.

substances. In fact, they have been able to measure the effectiveness of these substances by a standard bioassay comparison of the rates of growth. For this purpose discs of secondary phloem parenchyma of standard size are taken from the roots of selected strains of carrot and planted on a basic nutrient medium supplemented by liquid endosperm of a fixed concentration or by an extract of endosperm in a milk stage. Growth is improved also by the addition of casein hydrolysate.

Until now, most of the studies have been carried out on the milk or endosperm in the free nucleate stages of coconut fruits. Other investigations, however, have been made on extracts of immature corn grains in the milk stage and immature horsechestnut fruits. It is now clear that these preparations yield a variety of compounds, some more directly associated with the stimulation of cell enlargement than others. Substances, often chemically quite unrelated, tend to promote cell division. As examples, Steward reports identifying 1,3-diphenylurea in coconut milk and certain leucoanthocyanins in all endosperms studied. A new indole-acetic arabinose compound has been obtained in a highly active and apparently chemically pure state [15]. When added in very low concentrations to an otherwise complete medium, these substances are followed by cell division in the explants. Their activity, however, is fully developed only in the presence of reduced nitrogen represented by amino acids or casein hydrolysate [15]. It is also known that a neutral component of the endosperms, present in quantity, helps to create the background in which these more active agents function. It is important to note that all of these endosperms, which nourish immature embryos in vivo, also have the ability to stimulate adult, inactive, isolated carrot cells to grow. A large number of substances, including a spectrum of amino acids, has been isolated in considerable amounts in the fractionation of such endosperms as coconut milk. The definitive properties of the milk seem to be associated on the one hand with the somewhat miscellaneous group of substances which stimulate cell division per se, these being active at very low dilution, and on the other hand with the more synergistic, neutral, high water-soluble substances which permit the full activity of these stimulants of cell division to be developed [15]. Among these water-soluble substances found in the neutral fraction are to be listed three hexitols—the sugar alcohol, sorbitol, which Steward finds constitutes about 20 per cent by dry weight of coconut milk; and two inositols, myo-inositol and scyllo-inositol, in small amounts.

It is indeed interesting and probably very significant that the use of coconut milk in sterile nutrient culture studies, as the most plentiful source of endosperm, has proved exceedingly beneficial. In the nutrient culture studies of diverse plant materials, e.g. in the culture of embryos, in the growth of single cells, or very small pieces of plant tissue from seed plants, coconut milk added to the medium can be conducive to excellent growth, and especially so when supplemented by casein hydrolysate. In any case, Steward's studies [12-14] indicated that, for the growth of carrot cells in culture, if one wishes them to continue growth beyond the few-cell stage into young growing plants, it is necessary, in our present knowledge of nutrition, to add the liquid endosperm of coconuts or some equivalent endosperm. It is significant that, in the culture of the embryos or the parts of vascular plants other than the angiosperms, coconut milk proves unnecessary.

Generalizations on Early Morphogenesis

Certain seeming facts stand out as a result of the first of these studies on the growth of free plant cells now being carried on in different laboratories: *1*) Whole plants can originate from free cells derived from a callus mass, if judiciously nourished. It appears established that these free, parenchymatous cells are totipotent. *2*) The steps by which a single cell reaches that size of multicellular mass at which organogeny seems certain are not in any sense uniform. Multiple patterns in the orderliness of cell division play little but incidental part in the origin of the multicellular mass. The polarization of the plane of division of the first mitosis is not determined, nor is the timing of this division as to its occurrence before extensive growth has taken place. In fact, the regular sequence of cytokinesis following mitosis does not necessarily play a part even in a statistical statement of what constitutes the most frequent method of achieving the cell-mass stage. *3*) With the appearance of a radially multilayered cell mass, a nest or nodule of cells becomes centrally evident. The number of cells in this nodule increases as a result of periclinal divisions in the cambial-like layer surrounding or within it. Xylem elements are formed almost immediately and phloem may develop as well peripheral to the xylem in the nodule. This has not yet been tested. The formation of the endogenous group of vascular elements is frequently, and possibly always, followed by the development of an associated root which quickly becomes protuberant. This may then be followed by a shoot in a plane 180° away (i.e. diametrically opposite to the first root). The plant pattern with its root-shoot axis is thereby laid down and its further development is the stepwise procedure of that plant growing from a seedling. *4*) The pattern followed by those cells has highlights which compare favorably with those of developing embryos of seed plants from fertilized eggs, e.g. the filamentous early embryos of angiosperms, the free-nuclear stage of early embryos of gymnosperms, and so on. They are similar also in the same respects to natural and sequential stages in the development of a fern embryo, a club moss embryo, or one from Equisetum, or the young bud described by Sterling [16] in its growth from a nodule-like center of tobacco pith callus.

In substance therefore it appears that, in the sequence of visible and recognizable developmental stages, whether from a fertilized egg, or a free, vegetative cell, or from an embryonic bud, common biochemical problems must be solved in the production of an integrated organism. The early stages of origin, whether of vegetatively— or sexually—produced embryonic plants, are clearly not fixed or uniform. As pointed out earlier, they may arise in a variety of ways. Having achieved the three-dimensional stage of a cell aggregate by whatever means, subsequent development follows a more uniform pattern. But in the progressive unfolding of this pattern, whether from vegetative cell or fertilized egg, certain distinguishable and salient stages or "check points" may be observed. These check points are recognizable by common criteria and characteristics; e.g. the early more or less globular embryonic plant, the differentiation of centrally located vascular tissues, the origin of roots, and the origin of shoots with leaf primordia. What the biochemical criteria are that serve to define these seemingly critical stages or check points remains to be seen.

Early Morphogenesis in Ferns

A second morphogenetic problem follows which has proved to be especially interesting. To approach it, I shall refer to the life history of a fern, although a club moss or horsetail would serve as well. In the life cycle of a fern, two germs or single-celled initials are recognized, one for each phase of the life history, namely, the spore and the fertilized egg. It is common knowledge that the fertilized egg develops into a new plant, the plant which we designate as the fern. In fact, we recognize the different species of ferns by the phenotypic appearance of the fern plants derived from their respective fertilized eggs. By contrast, the spore grows into a small, flattish, green, thin plant structure generally described as a prothallus. This prothallus bears the sex organs, without which neither eggs nor sperms are developed. In the presence of moisture, the sex organs burst. Sooner or later, a free-swimming sperm effects the fertilization of an egg in its recently, often explosively opened container or archegonium.

It seems almost superfluous to state that the prothallus is as much a fern genetically as is that fern plant derived from the fertilized egg. They have the same genome. While it is true that the prothallus is haploid and the fertilized egg and its resulting fern plant are diploid, it is now common experience to the biologist that polyploidy is not by itself a sine qua non for phenotypic difference [17]. Haploid Jimson weeds and haploid corn plants, to mention certain well recognized examples, are known, which resemble diploid, tetraploid, or other polyploid plants of the same species in all respects except for possible minor differences in size. Of added interest are now known diploid and even tetraploid fern prothalli that have been produced in our laboratories and which closely resemble the haploid plants of the same species. It seems necessary on the face of it to search further than the contrast in chromosome number when one seeks to understand the contrastive nature of the two kinds of autophytic fern plants, the prothallus or sex organ-producing plant and the so-called fern or spore-producing plant. A striking difference comes into sight when one realizes that spores are shed and, if they germinate in nature, they do so wherever they fall, for they are nonmotile. By contrast, the egg is fertilized within a flask-shaped sac or archegonium and must develop there under conditions imposed by this organ and its surrounding tissues. In other words, the spore-to-prothallus stage develops on the ground, growing in unrestricted fashion; the resulting plant is essentially a two-dimensional structure, thin and green. When young, it is small and often achieves a heart shape, with its growing apex at the notch of the heart. If its early, ripe archegonia fail to be fertilized, it may become strap-shaped, it may branch, and it may put out prothallial lobes from its margins or even from its surfaces and become very irregular; in other words, it seems to grow all over. It may grow on, year after year, unless it dies from desiccation or unless it is stopped by fertilization and the development of a young embryo in one of its successive crops of archegonia.

By contrast, the fertilized egg is housed in the venter or base of the archegonium which develops in turn from the tissue of the prothallus, with only the neck protruding. Strikingly enough, the first visible effect of fertilization is not found in the egg. In the polypodiaceous fern upon which he worked, Ward [18] found the first recognizable aftermath of fertilization in the multiplication of the jacket cells of the archegonium, the divisions always taking place parallel to the surface of the venter in which the egg lies. More than five days intervened between the penetration of the sperm into the open neck of the archegonium and the first division of the fertilized egg. In this five-day period, the jacket can increase from two to four cells thick. The fertilized egg then begins its recognizable development as an embryo in its flask-shaped container, the calyptra. Under these conditions the egg follows an orderly sequence of cell divisions, early filling the space inside the thickening encasement of the venter walls. As elsewhere, so here, it is difficult to say with any degree of certainty that the planes of cell division of the young embryo are determined by or correlated with the restraint due to progressively more numerous, turgid, multiple-layered cells of the calyptra. Although difficult to demonstrate, it seems to be more than coincidence that under these conditions the sum total of growth of the restrained embryo here is the early formation of a nearly spherical entity. While indications of a more or less regular sequence can be found for the first three cell divisions, signs of any further uniformity are debatable; nor can one see in the quadrants or octants any sign of predictable organ formation. Almost half of the young embryo becomes an absorbing foot, composed of large cells which retain their close adherence to the cells of the surrounding prothallus. By about the tenth day, within the other half of the embryo, the cells in a central region take on a somewhat procambial character of elongate, narrow cells. At about this time, the first leaf is initiated by superficial cell divisions in the anterior part of the lower hemisphere. Elongating cells, which portend the procambial stage of the leaf trace, extend toward the leaf primordium almost simultaneously from the central nest of procambial cells. Within another two days the endogenous origin of a root is evident at a point in the lower half of the embryo and opposite to that of leaf initiation. Shortly therafter, within the next day or two, procambial cells extend from the central area to the incipient root. Growth of the root early assumes a faster rate than that of the leaf so that, for a time, it becomes the longer and larger organ. Interestingly enough, the stem is the last of the organs to appear, its initials becoming recognizable between the first leaf and the foot. Its vascular system is, from its initiation, continuous with that of the leaf and that of the central region of procambial tissue. By this time the young embryo is oriented transversely to the axis of the archegonium. Its developing leaf, stem, and root soon break through the enclosing calyptra. The embryo, still attached to the prothallus by an expanded foot, is dependent upon it for food. Its root early becomes established in the ground, its stem and leaf grow upward, tropistically oriented, and an independent, photosynthetic plant is established.

Growth of the embryo in the archegonium is therefore essentially an over-all or three-dimensional growth. From the nearly spherical young embryo, early centers can be recognized at one pole giving rise to the embryonic leaf and young stem and, at the other, to the first root. These become provided with extensions from the central provascular aggregation of cells so early that their organization is essentially predicated at the time of their origin. Such aspects of timing must become important in considering the physiology of the origins of parts.

I cannot refrain at this time from calling attention again to the contrast in the sequence of events in the spore development to a prothallus. Here, it seems, is a less restrained pattern of growth, the major developments being from the margins, and, in later stages of development, from outgrowths produced anywhere from the surface.

Experimental Approaches to Morphogenetic Problems

In endeavoring to bring the contrastive growths of the two propagating cells, the spore and the fertilized egg, under experimental control, some method of altering the conditions of development by changing the environment seemed logical. In his early efforts to remove the fern egg from the archegonium, after fertilization but before the first cell division, Ward [19] encountered difficulties, for the egg proved to be attached directly to the bottom wall of the venter. He then did the next best thing—he made clean cuts, either vertical or tangential or both, through the prothallus, as close as possible to the cavity of the venter without opening the cavity itself. The pad of prothallial cells below the archegonium served as a tissue through which the young embryo was nourished. If this isolated segment of prothallial tissue were provided with a known satisfactory nutrient medium, the embryo grew well. But, it did not develop in the regular, direct fashion that it followed when it was contained in an archegonium. Freed from its protective and restraining envelope of cells, the calyptra, divisions in the embryo were more haphazard and less predictable. When lateral and tangential cuts were made, the embryo tended to show more irregular divisions, burst early out of the venter walls, and appeared cylindrical rather than spherical. Its rate of development was retarded, leaves were slower in appearing but, when apparent, they proved to have veins of vascular tissue which were connected with a basal, internal nest of procambial cells. The initial leaf developed a stem that was associated with it, both connected with the basal vascular nest of cells. Thus, the cylindrical embryo produced a leaf or leaves, soon recognized in connection with an embryonic stem. No roots were developed in 60 days.

In those experiments in which only a tangential cut was made, thereby removing the neck of the archegonium and the outer calyptra tissues, startlingly different results followed. Early protrusion of the embryo from the now less restraining outer part of the venter was of an irregular and massive nature. The first appendages were more like fleshy protuberances than leaves, but they did have vascular connections with procambial cells in the more massive embryonic aggregation. Also, with each leaf there was associated an embryonic stem in vascular continuity with the central reservoir of vascular cells. The resulting young fern plant was more bushlike in its aggregation of stems, each with one or more leaves. Attachment of the foot to a part of the prothallus maintained nourishment for the embryo. Such slow growing massive embryos must be kept under longer observation in order to understand the relation of restraint and of nutrition to delayed leaf and root development.

Time does not permit further development of this study here. Suffice it to say that accumulating evidence suggests that it is not only the potentialities of a fertilized egg as opposed to those of a spore that determine the developmental pattern. Rather, the environment itself, physical as well as chemical, may so alter the early pattern that the end products are scarcely recognizable as the same plant species.* Much

* More recent pertinent work is reported by DeMaggio and Wetmore: *Am. J. Botany*, **48**, 551 (1961).

more needs to be done before one can envision the effects of different environmental variables on development. One might well ask if and how a single vegetative fern cell would grow if it could be put into an archegonium to develop instead of an egg. In fact, it would be even more striking if one could transplant a spore to it and have it grow.

It does seem reasonable to expect that experimental procedures can be invoked that should lead to a better understanding of the growth of embryos of angiosperms. Moreover, a great possibility is in sight for making a reality of embryo culture and therefore providing a better chance for plant breeders to preserve unusual hybrids that have died, until now, when left to develop in situ, but might well live when dissected out and transferred to a synthetic nutrient medium. Some results already exist in this field, enough to suggest that more work should be done.

Later Phases of Morphogenesis

A third set of morphogenetic problems which appears in the early embryology of vascular plants is unique to plants and certainly sets them apart from the animal. The polarization of the young embryo somehow separates the opposing ends physiologically, so that they develop into shoots and roots, respectively. But even more important is the fact that the two poles of the early embryonic mass remain actively meristematic, forming the so-called apical meristems of shoot and root. As the cells of these regions divide and leave the products of their divisions behind them, the two apical meristems become farther separated from each other. These potentially continuing meristematic apices make the plant what it is, a fixed organism with its extending root becoming more buried in the ground, and its shoot system with its main and branched apices farther removed from the root. The second attribute of these apices, shoot and root, is in the production of appendages. At regular time intervals, other things being equal, new leaves are produced. Associated with each leaf of an angiosperm is a bud primordium destined, if and when it grows, to produce a branch. On the root, lateral roots are produced, also characteristically in regular positions. Thus, appendages in plants are not limited in number as they are in animals. They continue to be produced throughout life; the loss of one or a few appendages is of little consequence in the survival of the plant. With each branch, root, or shoot, there is an apical meristem, and with this there is a continuing embryology throughout the life of the branch, unless the apex is lost by accident or becomes reproductive. One can thus state that any branch of a shoot system demonstrates throughout a growing season a continuum, from dividing cells producing new appendages distally through progressive stages of growth and development of leaves and buds on the growing stem which bears them, to mature stems with fully differentiated tissues and bearing fully grown, mature, functioning leaves. Similar statements can be made of the progressive panorama of change between a root apex and the mature root so that, in any growing season, all intermediate stages are present at any one time.

Recent studies have shown for shoots and roots that, even speaking biochemically, the natural differentiation of the tissues behind the apex is completely dependent upon the presence of the apex. Jacobs [20] has beautifully shown, recently, that normal development in the apical regions of Coleus is not only predictable but quantitatively demonstrable. In this, the two variables prove to be a sugar and an auxin, the latter at least emanating from the apical regions, especially the young leaves. Even more recently, Torrey [21] has proved that the same two chemicals are significant in the differentiation of the vascular tissues of roots, and here the source of the auxin is the root apex.

In our own laboratories we have been interested in demonstrating the effectiveness of the apical region in bringing about differentiation of tissues. For this we have grafted a growing bud of lilac into a homogeneous parenchymatous callus grown from the cambium of the same species. Not only does such a graft grow, but it very effectively induces the formation of vascular tissues in an otherwise homogeneous parenchyma [22-24].

Recently it has become apparent that not only do the xylem and phloem components of the vascular tissues develop, but they conform generally to the topographical pattern found in stems. It has now become apparent that we do not need to graft an apex of lilac to bring about the differentiation of vascular tissues. Instead, agar containing the growth hormone indole-3-acetic acid and a sugar—we have used both sucrose and glucose—is equally effective not only in inducing the vascular tissues but also in provoking their arrangement in a pattern definitely comparable with that of the stem. A low concentration of the growth hormone will cause, as seen in transverse section, a small circle of vascular nodules to be formed; if a higher concentration is used, a larger circle of more widely spaced nodules appears. In fact, these nodules can become connected by a cambial layer composed of fascicular and interfascicular portions as in the stem. Moreover, each of these nodular vascular entities has phloem externally and xylem internally as in a stem of lilac (Wetmore and Rier, in preparation). It seems fair to state that, with an apex of a lilac bud, or even with chemicals which are known to be produced by lilacs, as by vascular plants generally, one is able to implant upon a piece of parenchymatous callus a pattern of organization which suggests that the callus has now simulated a stem. There is a ring of vascular nodules surrounding a pseudopith and surrounded by parenchyma which, by no great stretch of the imagination, could be called cortex. More separated, isolated nodules may appear with central xylem completely encircled by phloem; they often have a cambium-like layer around the whole.

In this demonstration of chemical variables involved in the differentiation of vascular tissues, it begins to be apparent that, morphogenetically, we can now approach problems underlying the physiological processes involved. We have no answers as yet to explain how the same agents induce the formation of both xylem and phloem elements; experimentally it is apparent that a higher concentration of sugar is necessary for the latter than for the former. This is surprising since, in the differentiation of xylem, the conducting elements, whether tracheids or vessel elements, are not only provided with secondary cellulose walls but, in addition, have deposited microfibrils of lignin in these secondary walls between those of cellulose. No secondary walls occur in most conducting cells of phloem, and even in those members of the Pine family where they do exist they are not lignified [25]. Moreover, the role of auxin in the differentiation of either xylem or phloem is not at all clear biochemically. But, it would seem that a new approach may be provided for attacking these problems.

Where does all of this leave us? Clearly one can see organization in the developing as well as in the mature organism, the vascular plant. Though there is much of circumstantial nature that can be and must be checked experimentally, the small amount of direct evidence indicates that the early stages of the plant do not represent a rigid genetic pattern of development. Rather, we are forced to conclude that, for the stages between the single cell and the massive cell aggregate, factors, of which we now know little, determine the alternative sequences of pattern changes, pre-eminently changes relating to the plane of, and variation in, orderly cell division. We have some indication that contained cells, found naturally as fertilized eggs, act differently in their partially confining and probably even definitely restraining space than when germinating out in the open (DeMaggio and Wetmore; see footnote p. 231). When cultures of carrot callus can give isolated, isodiametric, small cells, or large cells which become multinucleate, or segmented filaments, or budded cells, all together under similar conditions of nutrition, it seems that at least some of the factors involved in early embryogeny or development must be microbiochemical in nature. What these seemingly obscure factors are is without any evidence at present. Do microchanges in pH or CO_2/O_2 balance, and so on, enter into these considerations?

For the moment, when the young entity, by whatever method, has produced a cell mass of several layers in thickness, or, stated differently, has attained in size a recognizable inside and outside, we can say that organizational pattern has been achieved. Planes of cell division in the central part of the colony seem to be polarized, and differentiation occurs in centrally placed cells which are capable of becoming xylem and phloem elements. If then, cultures are stabilized—no longer rotated—in an agar medium, one can follow the development of leaf and stem primordia and even of roots, each with characteristic vascular tissue. Apical meristems are ordained in root and shoot and the plant takes form. If this fundamental plan of origin from single cells (whether fertilized egg or vegetative, totipotent cell, whether free or in the plant) is a fair estimate of available information, we should be able to investigate the problems raised, both from the point of view of early, direct correlative effect of the physical environment and of later, direct endogenous biochemical changes. In any case, in morphogenetic studies of development, it now appears necessary to recognize that embryonic development does not proceed by a single, rigid, orderly pattern; rather, alternative patterns, each orderly, may occur side by side. The phenotypic events, however varied, are all within the genetic expression of the embryonic cell. Although we know at present almost nothing of the interpretation, much less the control, of these early stages, we must give our most earnest attention to cell environment. We must, at the same time, become more familiar with the reactions of the cell to its biophysical as well as its biochemical milieu before we can hope to comprehend such an all important morphogenetic problem as the orientation of the planes of cell division. While the biochemical properties of the medium may indicate the ability of cells to grow, their immediate physical environment, such as the embryo sac or the archegonium, may profoundly alter the way in which this growth is expressed.

REFERENCES

1. BALL, E. Development in sterile culture of stem tips and subjacent regions of *Tropaeolum majus* L. and of *Lupinus albus* L. *Am. J. Botany*, **33,** 301 (1946).
2. CAPLIN, S. M., and F. C. STEWARD. A technique for the controlled growth of excised plant tissue in liquid media under aseptic conditions. *Nature*, **163,** 920 (1949).
3. GAUTHERET, R. J. Recherches sur la culture des tissus végétaux: Essais de culture de quelque tissus méristématiques. *Rev. cytol. et cytophysiol. végétales*, Mem. 1, 279 pp. (1935).
4. GAUTHERET, R. J. La culture des tissus végétaux. Principes généraux. Nutrition. Applications à la pathologie. *La Nature*, pp. 364-67 (1952); 22-26, 53-57 (1953).
5. GAUTHERET, R. J. La culture des tissus végétaux: Physiologie et nutrition. *Actes soc. helv. sci. nat.*, pp. 40-52 (1955).
6. MOREL, G. Recherches sur la culture associée de parasites obligatoires et de tissus végétaux. *Ann. épiphyt.*, **14,** 1 (1948).
7. MILLER, C. O., and F. SKOOG. Chemical control of bud formation in tobacco stem segments. *Am. J. Botany*, **40,** 768 (1953).
8. SKOOG, F., and C. O. MILLER. Chemical regulation of growth and organ formation in plant tissues cultured *in vitro*. In: Symposium on Biological Action of Growth Substances. *Soc. Exp. Biol.*, **11,** 118 (1957).
9. MILLER, C. O., F. SKOOG, M. H. VON SALTZA, and F. M. STRONG. Kinetin, a cell division factor from deoxyribonueleic acid. *J. Am. Chem. Soc.*, **77,** 1392 (1955).
10. MILLER, C. O., F. SKOOG, F. S. OKUMURA, M. H. VON SALTZA, and F. M. STRONG. Structure and synthesis of kinetin. *J. Am. Chem. Soc.*, **77,** 2662 (1955).

11. MILLER, C. O., F. SKOOG, F. S. OKUMURA, M. H. VON SALTZA, and F. M. STRONG. Isolation, structure and synthesis of kinetin, a substance promoting cell division. *J. Am. Chem. Soc.*, **78,** 1375 (1956).
12. STEWARD, F. C., M. MAPES, and J. SMITH. Growth and organized development of cultured cells. I. Growth and division of freely suspended cells. *Am. J. Botany*, **45,** 693 (1958).
13. STEWARD, F. C., M. MAPES, and K. MEARS. Growth and organized development of cultured cells. II. Organization in cultures grown from freely suspended cells. *Am. J. Botany*, **45,** 705 (1958).
14. STEWARD, F. C. Growth and organized development of cultured cells. III. Interpretation of the growth from free cell to carrot plant. *Am. J. Botany*, **45,** 709 (1958).
15. STEWARD, F. C., and H. Y. MOHAN RAM. Determining factors in cell growth: Some implications for morphogenesis in plants. Chap. 5 in: *Advances in Morphogenesis*, vol. 1, pp. 189–265. New York, Academic Press (1961).
16. STERLING, C. Histogenesis in tobacco stem segments cultured *in vitro*. *Am. J. Botany*, **37,** 464 (1950).
17. STEBBINS, G. L., Jr. *Variation and Evolution in Plants*. New York, Columbia Univ. Press (1950).
18. WARD, M. The development of the embryo of *Phlebodium aureum* J. Sm. *Phytomorphology*, **4,** 18 (1954).
19. WARD, M., and R. H. WETMORE. Experimental control of development in the embryo of the fern, *Phlebodium aureum* J. Sm. *Am. J. Botany*, **41,** 428 (1954).
20. JACOBS, W. P. A quantitative study of xylem development in the vegetative shoot apex of Coleus. *Am. J. Botany*, **44,** 823 (1957).
21. TORREY, J. G. Experimental modification of development in the root. In: *Cell, Organism and Milieu*, D. Rudnick, Ed., pp. 189–222. New York, Ronald Press (1959).
22. WETMORE, R. H. The use of *"in vitro"* cultures in the investigation of growth and differentiation in vascular plants. Brookhaven Symposia in Biology No. 6: *Abnormal and Pathological Plant Growth*, pp. 22–40 (1954).
23. WETMORE, R. H., and S. SOROKIN. On the differentiation of xylem. *J. Arnold Arboretum (Harvard Univ.)*, **36,** 305 (1955).
24. WETMORE, R. H. Growth and development in the shoot systems of plants. In: *Cellular Mechanisms in Differentiation and Growth*, D. Rudnick, Ed., pp. 173–90. Princeton, Princeton Univ. Press (1956).
25. ABBE, L. B., and A. S. CRAFTS. Phloem of white pine and the other coniferous species. *Botan. Gaz.*, **100,** 695 (1939).

THE DEVELOPMENT OF LEARNING IN THE RHESUS MONKEY

By HARRY F. HARLOW
University of Wisconsin

DURING THE LAST five years we have conducted an integrated series of researches tracing and analyzing the learning capabilities of rhesus monkeys from birth to intellectual maturity. Control over the monkey's environment has been achieved by separating the infants from their mothers at birth and raising them independently, using techniques and methods adapted from those described by van Wagenen [1].

There are many characteristics that commend the rhesus monkey as a subject for investigation of the development of learning. At birth, or a few days later, this animal attains adequate control over its head, trunk, arm, and leg movements, permitting objective recording of precise responses on tests of learning. The rhesus monkey has broad learning abilities, and even the neonatal monkey rapidly learns problems appropriate to its maturational status. As it grows older this monkey can master a relatively unlimited range of problems suitable for measuring intellectual maturation. Although the rhesus monkey matures more rapidly than the human being, the time allotted for assessing its developing learning capabilities is measured in terms of years—not days, weeks, or months, as is true with most subprimate forms. During this time a high degree of control can be maintained over all experimental variables, particularly those relating to the animal's learning experiences. Thus, we can assess for all learning problems the relative importance of nativistic and experiential variables, determine the age at which problems of any level of difficulty can first be solved, and measure the effects of introducing such learning problems to animals before or after this critical period appears. Furthermore, the monkey may be used with impunity as a subject for discovering the effects of cerebral damage or insult, whether produced by mechanical intervention or by biochemical lesions.

The only other creature whose intellectual maturation has been studied with any degree of adequacy is the human child, and the data from this species attest to the fact that learning capability increases with age, particularly in the range and difficulty of learned tasks that can be mastered. Beyond this fact the human child has provided us with astonishingly little basic information on the nature or development of learning. Obviously, there are good and sufficient reasons for any and all such deficiencies. There are limits beyond which it is impossible or unjustifiable to use the child as an experimental subject. The education of groups of children cannot be hampered or delayed for purposes of experimental control over either environment or antecedent learning history. Unusual motivational conditions involving either deprivation or overstimulation are undesirable. Neurophysiological or biochemical studies involving or threatening physical injury are unthinkable.

Even aside from these cultural limitations, the human child has certain characteristics that render him a relatively limited subject for the experimental analysis of the maturation of learning capability. At birth his neuromuscular systems are so undeveloped that he is incapable of effecting the precise head, arm, hand, trunk, leg, and foot movements essential for objective measurement. By the time these motor functions have adequately matured, many psychological developmental-processes, including those involving learning, have appeared and been elaborated, but their history and nature have been obscured or lost in a maze of confounded variables.

By the time the normal child has matured physically, he is engaging each day in such a fantastic wealth of multiple learning activities that precise, independent control over any single learning process presents a task beyond objective realism. The multiple interactive transfer processes going on overwhelm description, and their independent experimental evaluation cannot be achieved. Even if it were proper to cage human children willfully, which it assuredly is not, this very act would in all probability render the children abnormal and untestable and again leave us with an insuperable problem.

It might appear that all these difficulties could be overcome best by studying the development of learning abilities in infraprimate organisms rather than monkeys. Unfortunately, the few researches which have been completed indicate that this is not true. Animals below the primate order are intellectually limited compared with monkeys, so that they learn the same problems more slowly and are incapable of solving many problems that are relatively easily mastered by monkeys. Horses and rats, and even cats and dogs, can solve only a limited repertoire of learning tasks, and they learn so slowly on all but the simplest of these that they pass from infancy to maturity before their intellectual measurement can be completed. Even so, we possess scattered information within this area. We know that cats perform more adequately on the Hamilton perseverance test than do kittens and that the same relationship holds for dogs compared with puppies [2,3]. It has been demonstrated that mature and aged rats are no more proficient on a multiple-T maze than young rats [4], and that conditioned responses cannot be established in dogs before 18–21 days of age [5]; but such data will never give us insight into the fundamental laws of learning or maturation of learning.

Neonatal and Early Infantile Learning: the First Sixty Days

Because learning and the development of learning are continuous, orderly processes, classifying learning into temporal intervals is an arbitrary procedure. However, a criterion that may be taken for separating early learning from later learning is the underlying motive or incentive. Solid foods are precluded as incentives for monkey learning prior to 40–60 days of age, forcing the experimenter to depend upon such rewards as liquid nutrients, shock avoidance, exploration, and home cage conditions. It is recognized that these same rewards may be used to motivate older primates on learning tasks, but for them the convenient incentive of solid food becomes available. Another arbitrary criterion that may be taken for choosing this temporal period lies in the fact that fear of strange, new situations—including test situations—only appears toward the end of this period.

Conditioned responses. The earliest unequivocal learned responses that we obtained from the rhesus monkey were conditioned responses in a situation in which an auditory stimulus was paired with electric shock. The standard procedure was to adapt the neonatal monkeys during the first two days of life by placing them for 10 minutes a day in the apparatus, which consisted of a cubic Plexiglas stabilimeter with a grid floor, enclosed in a sound-deadened cabinet with a one-way-vision screen on the front (see Fig. 40). Conditioning trials were initiated on the third day, the tone and shock intervals being mechanically fixed at 2 seconds and 1 second, respectively, and administered either separately or paired.

FIG. 40. Neonatal monkey in stabilimeter.

The animals were divided into three groups and were given daily trials as follows: five experimental subjects (T-S group) were given eight paired tone-shock trials and two test trials; four pseudoconditioning controls (P-C group) were given eight shock trials and two test trials in which tone only was presented; and four stimulus-sensitization controls (T-O group) received ten tone trials but never received shock from the grid floor. Conditioned and unconditioned responses were measured in terms of both the continuous, objective activity records taken from an Esterline-Angus recorder and the check-list records made by two independent human observers.

The learning data presented in Figure 41 show early and progressive learning. The differences between the frequency of conditioned responses by the five experimental subjects and the four subjects in each of the control groups were

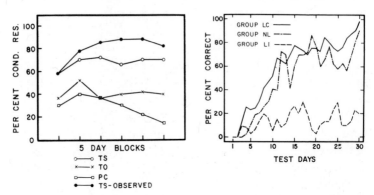

Fig. 41. Conditioned responses to tone.

Fig. 42. Straight-runway performance.

significant, even though clear-cut evidence of pseudoconditioning was found in one of the P-C animals. It will be noted that the observers recorded a higher frequency of conditioned responses than could be identified from the stabilimeter record. The observational data indicate that these tone-shock conditioned responses were learned by three subjects on the second test day and that unequivocal conditioning took place in four of the five subjects. The observational data also show that the form of the conditioned response changes with training, starting as a diffuse response and gradually becoming more precise. As training progressed, most subjects responded to the conditioned stimulus by standing erect, sometimes on one foot.

Limited tests failed to demonstrate any generalization of the conditioned response to the experimenter or to auditory stimuli presented outside the test situation. Retention tests made 15 days after the completion of the original training revealed very considerable learning loss, ranging among individual subjects from no definite indication of retention, to conditioned responses on about half the test trials.

Straight-runway performance. An apparently simple learned response, which has been frequently used by psychologists in studying rat learning, is the straight runway. We produced such an apparatus by simply using the monkeys' living cage as the runway and introducing a nursing booth at one end prior to each feeding period. At the time of testing the subject was taken to the far end of the home cage, faced toward the nursing booth, released, and allowed 30 seconds to enter the booth. The number of daily trials was determined by number of feeding sessions, twelve a day during the first two weeks, and ten a day subsequently.

The subjects were divided into three groups. For the light-conditioned animals (L-C group) the nursing booth was suffused with flashing green light during each of the training trials; for the no-light monkeys (N-L group) there were no conditioning cues other than those afforded by the test situation and the act of orientation; for the light-extinguished subjects (L-I group) the nursing booth was suffused with green light, but this light was immediately extinguished when the monkey entered the booth and simultaneously there began a 5-minute delay period before feeding.

The data presented in Figure 42 offer evidence of rapid and progressive improvement in performance. Many of the failures during the first 10 days resulted from locomotor limitations or from the disturbing effects of reorientation and restraint by the experimenter. It is clear, however, that learning occurred early in life and that the cue of green light added little or nothing to the cues provided by the presence of the experimenter and postural orientation. The L-I group, which did not receive food upon approach to the nursing booth, was significantly inferior to the other two groups, and it is possible that the green light became a cue for absence or delay of feeding.

Spatial discrimination. Two groups of ten monkeys each were tested on a spatial discrimination problem requiring choice of the right or left alley of the Y-maze illustrated in Figure 43.

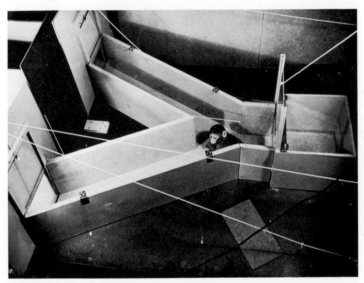

FIG. 43. Infant monkey Y-maze.

One group of subjects began training at 15 days of age (group 15), after 4 days of adaptation, and the other group started maze learning at 45 days of age (group 45). Two trials were given each day, a correct trial being rewarded by entrance into the home cage, a highly effective incentive for the infant monkey, whereas an incorrect response, defined as entrance into the incorrect antechamber, was punished by a 1-minute delay before rerunning. A rerun technique was used throughout this test; i.e., whenever the monkey made an error, it was returned to the starting position and run again until it made the correct choice and reached the home cage. Spatial discrimination learning was continued for 25 days; on day 26 the position of the correct goal box was reversed and the same training schedule of two trials per day continued.

The percentages of correct initial responses made by group 15 on days 1, 2, 5, 10, and 15 are 45, 60, 75, 75, and 95, respectively. Comparable percentages for group 45 are 80,

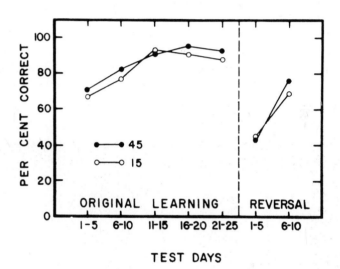

FIG. 44. Per cent correct responses on Y-maze.

55, 65, 85, and 100. Despite the high percentage of correct responses made by group 45 on day 1, the two learning curves, as illustrated in Figure 44, are very similar. Excluding a single member in each group that failed to adapt to the test situation and never met the criterion of eighteen correct responses in twenty consecutive trials, the mean number of trials to this criterion, excluding the criterional trials, was 8.5 for group 15 and 6.2 for group 45.

The percentage of correct responses dropped below chance

for both groups of monkeys during the first five reversal trials, and trial 1 was especially characterized by multiple, persistent, erroneous choices. During all these trials the animals made many violent emotional responses as indicated by balking, vocalization, and autonomic responses, including blushing, urination, and defecation. Even so, all but one subject in group 15 attained the criterion of eighteen correct responses in twenty consecutive trials, and the mean number of trials to learn, not including the criterional trials, was 19.2 for group 15 and 11.9 for group 45.

Although the performance of the older group was superior to that of the younger, particularly on the reversal problem, the differences were not statistically significant. Certainly the 15-day-old macaques solved this spatial learning task with facility, and their performance leaves little to be gained through additional maturation.

FIG. 45. Black-white discrimination apparatus.

Object discrimination. Two groups of four newborn monkeys were trained on a black-white discrimination—i.e. a non-spatial or object discrimination—by teaching them to select and climb up a black or white ramp for the reward of a full meal delivered through a nursing bottle. An incorrect choice was punished by a 3-minute delay in feeding. As can be seen in Figure 45, not only the ramp, but the entire half of the test situation, was black or white, as the case might be, and the positions of these half-cages were reversed on 50 per cent of the trials. The number of test trials was stabilized at nine per day after the first few days of life.

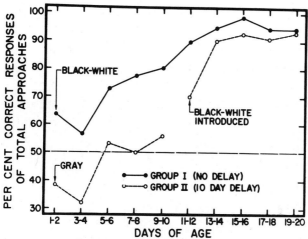

FIG 46. Black-white discrimination learning.

Learning by the neonatal macaques in this test situation proved to be almost unbelievably rapid, even allowing for the fact that a maximally efficient stimulus display was provided by the totally black and totally white halves of the test chamber. As can be seen in Figure 46, the group of infants trained from birth on the object-discrimination problem attained the criterion of 90 per cent correct responses on 2 consecutive days beginning at 9 days of age. This was a total of less than 100 trials, many of which were failed through physical inability to climb the ramp. A second group, run as a maturational control, was rewarded for climbing up either of two gray ramps for the first 10 days. On day 11 the black and white ramps were substituted, and these monkeys solved the black-white discrimination problem, the first formal learning problem they had ever faced, by the second test day, averaging less than thirteen trials to achieve the criterion.

After the black-white discrimination problem was solved, the infants were tested on discrimination reversal; i.e. the color of ramp previously correct was now made incorrect, and the color of the ramp previously incorrect became correct. The results were very similar to those obtained in the spatial discrimination problem. The infants made a great many errors when the problem was first reversed and showed very severe emotional disturbances. This was particularly true when the reversal went from white correct to black, since infant monkeys strongly prefer white to black.

We have also a considerable body of data showing that the infant monkey can solve form discriminations and color discriminations as well as the black-white brightness-discrimination problem. It is not possible in these other situations, particularly in the case of form discrimination, to attain the maximally efficient stimulus display previously described. For this reason—and a control study suggests that it is for this reason alone—the number of trials required to learn increases and the age at which learning can be demonstrated also advances. Even so, it has been possible to obtain discrimination between a triangle and a circle by the 20- to 30-day-old monkey after less than 200 training trials.

Infant Learning: The First Year

The most surprising finding relating to neonatal learning was the very early age at which simple learning tasks could be mastered. Indeed, learning of both the simple conditioned response and the straight runway appeared as early as the animal was capable of expressing it through the maturation of adequate skeletal motor responses. Thus, we can in no way exclude the possibility that the monkey at normal term, or even before normal term, is capable of forming simple associations.

Equally surprising is the fact that performance may reach or approach maximal facility within a brief period of time. The 5-day-old monkey forms conditioned reflexes between tone and shock as rapidly as the year-old or the adult monkey. The baby macaque solves the simple straight-alley problem as soon as it can walk, and there is neither reason nor leeway for the adult to do appreciably better. Although we do not know the minimal age for solution of the Y-maze, it is obviously under 15 days. Such data as we have on this problem indicate that the span between age of initial solution and the age of maximally efficient solution is brief. One object discrimination, the differentiation between the total-black and total-white field, shows characteristics similar to the learning already described. The developmental period for solution lies between 6 and 10 days of age, and a near maximal learning capability evolves rapidly. However, it would be a serious mistake to assume that any sharply defined critical periods characterize the development of more complex forms of learning or problem solving.

Object discrimination learning. Although the 11-day-old monkey can solve a total-black versus total-white discrimination problem in less than thirteen trials, the 20- to 30-day-old monkey may require from 150 to 200 trials to solve a triangle-circle discrimination problem when the stimuli are relatively small and placed some distance apart. It is a fact that, even though the capability of solving this more conventional type of object-discrimination problem exists at 20 days, object-discrimination learning capability has by no means attained full maturity at this time.

The development of complete object-discrimination capacity was measured by testing five different age groups of naive rhesus monkeys on a single discrimination problem. Discrimination training was begun when the animals were 60, 90, 120, 150, or 366 days of age, and, in all cases, training was preceded by at least 15 days of adaptation to the apparatus and to the eating of solid food. There were eight subjects in group 366 (as defined by age), ten in group 60, and fifteen in each of the other groups. A Wisconsin General Test Apparatus, illustrated in Figure 47, was used throughout the test sessions. A single pair of three-dimensional stimuli differing in multiple attributes such as color, form, size, and material was presented on a two-foodwell test tray of the Klüver type. The animals were given 25 trials a day, 5 days a week, for 4 weeks, a total of 500 trials. A noncorrection method was always used.

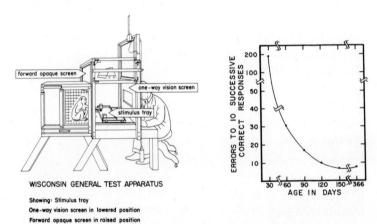

FIG. 47. Wisconsin General Test Apparatus.

FIG. 48. Initial discrimination learning as a function of age.

Figure 48 presents the number of trials taken by the five different groups of monkeys, and performance by a 30-day-old group on a triangle-circle discrimination is plotted on the far left. Whether or not one includes this group, it is apparent that the ability of infant monkeys to solve the object-discrimination problem increases with age as a negatively accelerated function and approaches or attains an asymptote at 120 to 150 days.

Detailed analyses have given us considerable insight into the processes involved in the maturation of this learning function. Regardless of age, the monkeys' initial responsiveness to the problem is not random or haphazard. Instead, almost all the subjects approached the problem in some systematic manner. About 20 per cent of the monkeys chose the correct object from the beginning and stayed with their choice, making no errors! Another 20 per cent showed a strong preference for the incorrect stimulus and made many errors. Initial preference for the left side and for the right side was about equally frequent, and consistent alternation patterns also appeared. The older, and presumably brighter, monkeys rapidly learned to abandon any incorrect response tendency. The younger, and presumably less intelligent, monkeys persisted longer with the inadequate response tendencies and very frequently shifted from one incorrect response tendency to another before finally solving the problem. Systematic responsiveness of this type was first described by Krechevsky [6] for rats and was given the name of "hypotheses." Although this term has unfortunate connotations, it was the rule and not the exception that our monkey subjects went from one "hypothesis" to another until solution, with either no random trials or occasionally a few random trials intervening. The total number of incorrect, systematic, response tendencies before problem solution was negatively correlated with age.

These data on the maturation of discrimination learning capability clearly demonstrate that there is no single day of age or narrow age-band at which object-discrimination learning abruptly matures. If the "critical period" hypothesis is to be entertained, one must think of two different critical periods, a period at approximately 20 days of age, when such problems can be solved if a relatively unlimited amount of training is provided, and a period at approximately 150 days of age, when a full adult level of ability has developed.

Delayed response. The delayed-response problem has challenged and intrigued psychologists ever since it was initially presented by Hunter [7]. In this problem the animal is first shown a food reward, which is then concealed within, or under, one of two identical containers during the delay period. The problem was originally believed to measure some high-level ideational ability or "representative factor"—a capacity that presumably transcended simple trial-and-error learning. Additional interest in the problem arose from the discovery by Jacobsen [8] that the ability to solve delayed-response problems was abolished or drastically impaired by bilateral frontal lobectomy in monkeys.

Scores of researches have been conducted on delayed-response problems. Almost all known laboratory species have been tested and all conceivable parameters investigated. Insofar as the delayed response is difficult, it appears to be less a function of period of delay or duration of memory than an intrinsic difficulty in responding attentively to an implicit or demonstrated reward. However, in spite of the importance of the problem and the vast literature which has accumulated, there has been no previous major attempt to trace its ontogenetic development in subhuman animals.

Ten subjects in each of four groups, a 60-, 90-, 120-, and 150-day group, were tested on so-called 0-second and 5-second delayed responses (the actual delay period is approximately 2 seconds longer) at the same time they began their discrimination learning. A block of ten trials at each delay interval was presented 5 days a week for 18 weeks, a total of 90 test days. These 900 trials at each delay interval constituted the test program for Series I, which was followed by

Series II during which time delay intervals of 5, 10, 20, and 40 seconds were introduced in counterbalanced order for 12 test weeks of 5 days each at the rate of 8 trials a day for each condition.

The results for the four infant groups on the 5-second delayed responses and the performance of a group of adults with extensive previous test experience on many different problems are presented in Figure 49. The four infant groups

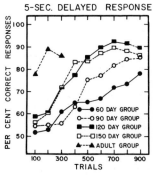
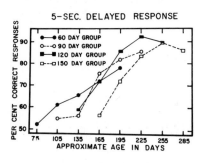

FIG. 49. Delayed-response learning as a function of age.

FIG. 50. Delayed-response performance of different maturational groups with age held constant.

show increasing ability to solve delayed responses both as a function of experience and as a function of age. The performance of all infant groups is inferior to that of the adult group, but differences in past learning experiences preclude any direct comparison.

The performance of the four infant groups of monkeys on the 5-second delayed responses for trials 1–100, 201–300, 401–500, 601–700, and 801–900 is plotted in Figure 50. Because performance during trials 1–100 is poor regardless of group, it is apparent that a certain minimum experience is required to master the delayed-response task. At the same time, the increasingly steep slopes of the learning curves make it apparent that efficiency of delayed-response learning and performance is in large part a function of age. The group data suggest that, after extensive training as provided in this study, 70 per cent correct responses may be attained by 150 days of age, 80 per cent by 200 days, and 90 per cent by 250 days.

The performance at the 40-second delay interval by all monkeys tested in Series II is presented in Figure 51. The

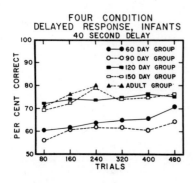

Fig. 51. Delayed-response generalization to 40-second interval.

performances of the three older groups and the two younger groups are similar, and there are no significant differences between the adult group and the two older infant groups. Similar results were obtained on the 5-, 10-, and 20-second delay intervals of the Series II tests except for the fact that the differences between the younger and older groups diminished progressively with decreasing delay intervals.

Very marked individual differences were disclosed during delayed-response testing, a finding which typifies this task regardless of species or age. Some monkeys, as well as some other animals, fail to adapt to the requirements of the test. Inspection of individual records reveals that the capability of solving this problem first appears at about 125 to 135 days of age and that essentially faultless performance may appear by 200 to 250 days in perhaps half the infant monkey subjects. Thus, some monkeys at this age may possess an adult capability, and these data are in keeping with the results obtained in the Series II tests. We have recently completed a study on a

30-month-old group of five monkeys on 0- and 5-second delayed responses, and their learning rates and terminal performance are at adult levels. Thus, it appears that we have definitive data on the maturation and acquisition of delayed-response performance by rhesus monkeys.

It is obvious that the capability of solving the delayed-response problem matures at a later date than the capacity of solving the object-discrimination problem. This is true regardless of the criterion taken, whether it is the age at which the task can be solved after a relatively unlimited number of trials or the age at which a full adult level of mastery is attained. At the same time, it should be emphasized that this capacity does develop when the monkey is still an infant, long before many complex problems can be efficiently attacked and mastered. Thus, there is no reason to believe that the delayed response is a special measure of intelligence or of any particular or unusual intellectual function.

Object discrimination learning set. The present writer, in 1949, demonstrated that adolescent or adult monkeys trained on a long series of six-trial discrimination problems showed progressive improvement from problem to problem [9]. As successive blocks of problems were run, the form of the learning curve changed from positively accelerated, to linear, to negatively accelerated; finally, there appeared to be two separate curves or functions; i.e. performance changed from chance on trial 1 to perfection or near perfection on and after trial 2. From trial 1 to 2 the curve is precipitate and from trial 2 onward it is flat. This phenomenon, called "learning set formation" or "interproblem learning," has proved to be a useful tool in comparative, physiological, and theoretical psychology. To obtain evidence concerning the maturational factors involved, the performance of various age groups of monkeys was measured on this task.

The same five infant groups previously tested on a single object-discrimination problem served as subjects for learning-set training. Upon completion of the original discrimination problem they were tested on four discrimination problems a day 5 days a week, each problem six trials in length. Group 366 was trained on 400 problems and the other monkeys on 600 problems. The individual test-trial procedures were identical with those employed in regular object-discrimination learning, but a new pair of stimuli was introduced for each new problem.

The trial 2 performance of the five groups of infant monkeys is plotted in Figure 52, and data from mature monkeys

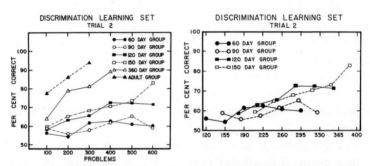

FIG. 52. Learning set formation as a function of age.

FIG. 53. Learning set plotted for age of completion of consecutive 100 problem blocks.

tested in previous experiments are also given. The two younger groups fail to respond consistently above a 60 per cent level even though they were approximately 10 and 11 months of age at the conclusion of training. The next two older groups show progressive, though extremely slow, improvement in their trial 2 performances, with groups 120 and 150 finally attaining a 70 and 80 per cent level of correct responding. These data are in general accord with those obtained from an earlier, preliminary experiment and indicate that the year-old monkey is capable of forming discrimination learning sets even though it has by no means attained an adult level of proficiency.

In Figure 53 the trial 2 learning-set performance for the various groups is plotted in terms of age of completion of consecutive 100-problem blocks, and these data suggest that the capacity of the two younger groups to form discrimination learning sets may have been impaired by their early, intensive learning-set training, initiated before they possessed any effective learning-set capability. Certainly, their performance from 260 days onward is inferior to that of the two groups with less experience but matched for age. The problem which these data illustrate has received little attention among experimental psychologists. There is a tendency to think of learning or training as intrinsically good and necessarily valuable to the organism. It is entirely possible, however, that training can be either helpful or harmful, depending upon the nature of the training and the organism's stage of development.

Because of the fundamental similarities existing between the learning of an individual problem and the learning of a series of problems of the same kind, it is a striking discovery that a great maturational gulf exists between efficient individual-problem learning and efficient learning-set formation. Information bearing on this problem has been obtained through the author's detailed error-factor analysis technique [10], which reveals that, with decreasing age, there is an increasing tendency to make stimulus-perseveration errors; i.e. if the initially chosen object is incorrect, the monkey has great difficulty in shifting to the correct object. Furthermore, with decreasing age there is an increasing tendency to make differential-cue errors—a difficulty in inhibiting, on any particular trial, the ambiguous reinforcement of the position of the stimulus which is concurrent with the reinforcement of the object per se.

In all probability, individual-problem learning involves elimination of the same error factors or the utilization of the same hypotheses or strategies as does learning-set formation.

However, as we have already seen, the young rhesus monkey's ability to suppress these error factors in individual-problem learning does not guarantee in any way whatsoever a capacity to transfer this information to the interproblem learning-set task. The learning of the infant is specific and fails to generalize from problem to problem, or, in Goldstein's terms, the infant possesses only the capacity for concrete thinking. Failure by the infant monkey to master learning-set problems is not surprising inasmuch as infraprimate animals, such as the rat and the cat, possess the most circum-

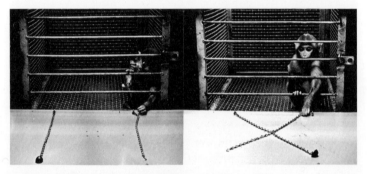

FIG. 54. Infant monkeys responding to the parallel pattern and the crossed-strings pattern.

scribed capabilities for these interproblem learnings, and it is doubtful if the pigeon possesses any such ability at all. Indeed, discrimination learning-set formation taxes the prowess of the human imbecile and apparently exceeds the capacity of the human idiot.

Most of the findings which we have reported for the maturation of learning in the rhesus monkey had been predicted, but this was not true in the case of the string tests. Two of the theoretically simple patterns, the parallel strings and the two crossed strings, are illustrated in Figure 54. We had assumed that the infant monkey would solve the parallel pattern with few or no errors, but, as can be seen in Figure 55, this assumption did not accord with fact. The infant monkeys made many errors, learned slowly, and in many cases failed to reach a level of perfect responsiveness after prolonged training. The data on the relatively simple two-crossed-strings pattern (Fig. 56) show that the 6-month-old rhesus monkey is just beginning to reach the age at which this problem can be mastered. Unfortunately, our data on patterned-strings learning are incomplete, but it is obvious that this capacity is a function which is maturing during the second half of the first year of life and probably for a considerable period of time henceforward. In retrospect, we

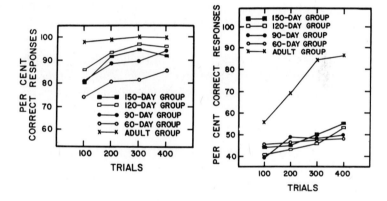

FIG. 55. Parallel string pattern performance as a function of age.

FIG. 56. Two-crossed-strings pattern performance as a function of age.

realized that the relatively late maturation of string-test learning was in keeping with known facts. The crossed-strings pattern has never been solved by any infraprimate animal, and this task cannot be resolved by the human infant [11, 12] until the second or third year of life.

The Development of Terminal Learning Ability

At the present time we have completed a series of experiments which clearly demonstrate that the capability of solving problems of increasing complexity develops in rhesus monkeys in a progressive and orderly manner throughout the first year of life. Furthermore, when we compare the performances of the year-old monkey and the adult monkey, it becomes obvious that maturation is far from complete at the end of the first year. Although our data on early development are more complete than our data on terminal learning capacities, we have already obtained a considerable body of information on middle and late learning growth.

Hamilton perseverance test. Just as we were surprised by the delayed appearance of the capability of mastering the patterned-strings tests, so were we surprised by the delay before performance on the Hamilton perseverance test attains maximal efficiency.

Three groups of monkeys were initially tested at 12, 30, and 50 months of age, respectively. The groups comprised six, five, and seven monkeys, and all were tested twenty trials a day for 30 days. On the perseverance problem the animal is faced with a series of four boxes having spring-loaded lids

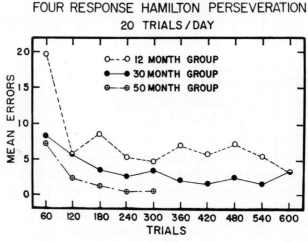

Fig. 57. Mean number of errors per 60-trial block on Hamilton perseverance test.

which close as soon as they are released. Only one box contains food, and the rewarded box is changed in a random manner from trial to trial with the provision that the same box is never rewarded twice in succession. In the present experiment the subjects were allowed only four responses per trial, whether or not the reward was obtained, and an error was defined as any additional response to an unrewarded box after the initial lifting of the lid during a trial. Infraprimate animals make many errors of this kind, but as can be seen in Figure 57, the mature monkey makes few such errors and learns rather rapidly to eliminate these. We were surprised by the inefficient performance of the year-old monkey and unprepared to discover that maximally efficient performance was not attained by the 30-month-old monkeys.

The mature monkey finds a simple plan for attacking the perseverance problem. Typically, it chooses the extreme left or right box and works systematically toward the other end.

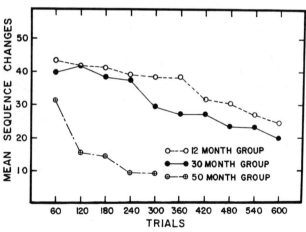

FIG. 58. Mean number of sequence changes per 60-trial block on Hamilton perseverance test.

If it adopts some more complex strategy, as responding by some such order as box 4–2–3–1, it will repeat this same order on successive trials.

Since the animal's procedural approach to the perseveration problem appeared to be an important variable, measures were taken of changes in the animal's order of responding from trial to trial, and these were defined as response-sequence changes. The data of Figure 58 show that the 50-month-old monkeys adopt the invariant type of behavior described above but that this is not true for either the 12- or 30-month-old groups. If a subject adopts an invariant response pattern, the problem is by definition simple; failure to adopt such a pattern can greatly complicate the task. In view of this fact it is not surprising that the 30-month-old subjects made so many errors; rather, it is surprising that they made so few—their error scores represent a triumph of memory over inadequate planning.

Relatively little research on the Hamilton perseverance method has been conducted by psychologists in spite of the fact that the original studies resulted in an effective ordering of Hamilton's wide range of subjects in terms of their position within the phyletic series. Furthermore, the limited ontogenetic material gave proper ordering of animal performance: kittens, puppies, and children were inferior to cats, dogs, and human adults. Above and beyond these facts, the perseverance data give support to the proposition that the rhesus monkey does not attain full intellectual status until the fourth or fifth year of life.

Oddity test. Probably the most efficient tests that have been developed for measuring the maximal intellectual capability of the subhuman primates are the multiple-sign tests, whose solution is dependent upon appropriate responses to multiple, simultaneously presented cues. One of the simplest of these tests is that of oddity. Each oddity problem utilizes two identical pairs of stimuli, the "A" and the "B" stimuli. On any trial, three of the stimuli are presented simultaneously, as is shown in Figure 59, and the odd or different stimulus is correct and rewarded.

FIG. 59. Correct response to oddity problem.

We have now completed a series of experiments using the two-position oddity problem, in which the correct stimulus is either in the right or left position, never in the center. In Figure 60 are presented data from a group of ten monkeys tested on 256 problems at 20 months of age and again at 36 months of age. Comparable data for a group of six adult rhesus monkeys are also graphed. These data indicate increasingly efficient performance as a function of age. The performance differences at each of the three age levels are statistically significant, and there is every reason to believe that intellectual maturation as measured by this test is incomplete at 3 years.

Additional oddity learning data obtained by the same training techniques are presented in Figure 61. Again a group of ten monkeys was trained on oddity, first at 12 and subsequently at 36 months of age. At 30 months of age, however, this group was divided into two groups of five monkeys each, one group being trained on a series of 480 six-trial discrimination problems and the other on 2,400 delayed-response trials. The differences between the two groups on the oddity problems are statistically significant, with every indication of negative transfer from the learning-set to the

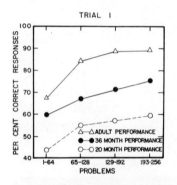

Fig. 60. Performance on oddity problems as a function of age.

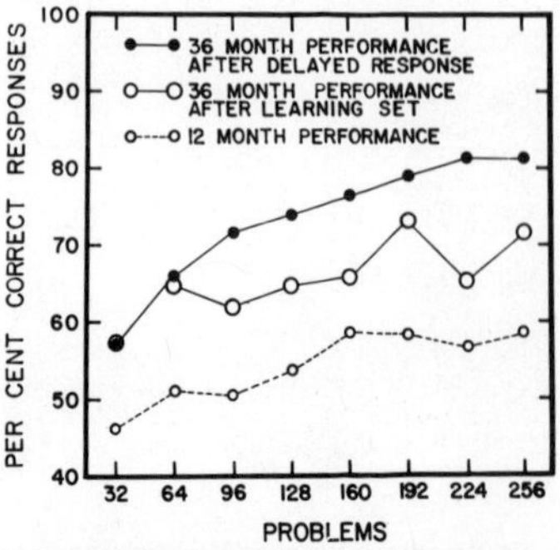

Fig. 61. Performance on oddity problems as a function of age and past experience.

oddity training. This is consistent with the fact that a single stimulus is uniformly correct on every discrimination problem but reverses frequently during the trials of each oddity problem. The transfer from delayed response to oddity may have been positive, since performance of the 36-month-old group with delayed-response training is superior to the comparable 36-month-old group whose performance is included in Figure 60. But, in neither case does a 36-month-old group attain an adult performance level.

For the neonatal and infant rhesus monkey each learning task is specific unto itself, and the animal's intellectual repertoire is composed of multiple separate and isolated learned experiences. With increasing age, problem isolation changes to problem generalization, and this fundamental reorganization of the monkey's intellectual world apparently begins in its second year of life. From here on, we can no longer specify the monkey's learning ability for any problem merely in terms of maturational age and individual differences. The variable of kind and amount of prior experience must now be given proper value. This is characteristic of monkey learning, and, in fact, learning by all higher primates as they come of intellectual age.

Monkeys do not attain full oddity performance at three years of age, and oddity learning is by no means the most complicated learning task that the adult rhesus monkey can solve. Oddity-nonoddity learning, in which the subject is required to choose either the odd or the nonodd stimulus, depending upon the color of the test tray presented on a particular trial, can be solved relatively routinely by the adult monkey. Considerably more complex learning problems have been mastered by highly trained rhesus subjects. We are in no position to determine the monkey's age of intellectual maturity, but four or five years of age is a reasonable estimate.

Summary and Interpretation

Half a decade is entirely too brief a period to establish a definitive program on the maturation of learning ability in the rhesus monkey, particularly in the later age ranges. However, within this period of time we have developed the techniques and conducted tests which demonstrate that such a program is entirely feasible. The monkey is capable of solving simple learning problems during the first few days of life and its capability of solving ever increasingly complex problems matures progressively, probably for four to five years.

Early in life, new learning abilities appear rather suddenly within the space of a few days, but, from late infancy onward, the appearance of new learning powers is characterized by developmental stages during which particular performances progressively improve. There is a time at which increasingly difficult problems can first be solved, and a considerably delayed period before they can be solved with full adult efficiency.

The monkey possesses learning capacities far in excess of those of any infraprimate animal, abilities probably comparable with those of low-level human imbeciles. The monkey's learning capabilities can give us little or no information concerning human language, and only incomplete information related to thinking. These are the generalizable limits of learning research on rhesus monkeys, but they still leave us with an animal having vast research potentialities. There is a wealth of learning problems which the monkey can master, and at present the field is incompletely explored. The maturation of any learning function can be traced, and the nature and mechanisms underlying interproblem and intertask transfer can be assessed. There exist great research potentialities in analyzing the fundamental similarities and differences among simple and complex learnings within a single species. The monkey is the subject ideally suited for studies involving neurological, biochemical, and pharmacological correlates of behavior. To date, such studies have been limited to adult monkeys, or monkeys of unspecified age, but such limited researches are no longer a necessity. We now know that rhesus monkeys can be raised under completely controlled conditions throughout a large part, and probably all, of their life span, and we may expect that the research of the future will correlate the neurophysiological variables, not with the behavior of the static monkey, but with the behavior of the monkey in terms of ontogenetic development.

REFERENCES

1. VAN WAGENEN, G. The monkey. In *The Care and Breeding of Laboratory Animals*, E. J. Farris, Ed., pp. 1–42. New York, Wiley (1950).
2. HAMILTON, G. V. A study in trial and error reactions in animals. *J. Animal Behavior*, **1,** 33 (1911).
3. HAMILTON, G. V. A study of perseverance reactions in primates and rodents. *Behav. Monogr.*, **3,** No. 2, 1 (1916).
4. STONE, C. P. The age factor in animal learning. II. Rats on a multiple light discrimination box and a difficult image. *Genet. Psychol. Monogr.*, **6,** No. 2, 125 (1929).
5. FULLER, J. F., C. A. EASLER, and E. M. BANKS. Formation of conditioned avoidance responses in young puppies. *Am. J. Physiol.*, **160,** 462 (1950).
6. KRECHEVSKY, I. "Hypothesis" vs. "chance" in the pre-solution period in sensory discrimination learning. *Univ. Calif. Publ. Psychol.*, **6,** 27 (1932).
7. HUNTER, W. F. The delayed reaction in animals and children. *Behav. Monogr.*, **2,** 21 (1913).
8. JACOBSEN, C. F. An experimental analysis of the frontal association areas in primates. *Arch. Nerv. Ment. Dis.*, **82,** 1 (1935).
9. HARLOW, H. F. The formation of learning sets. *Psychol. Rev.*, **56,** 51 (1949).
10. HARLOW, H. F. Analysis of discrimination learning by monkeys. *J. Exp. Psychol.*, **40,** 26 (1950).
11. MATHESON, E. A study of problem solving behavior in pre-school children. *Child Develop.*, **2,** No. 4, 242 (1931).
12. RICHARDSON, H. M. The growth of adaptive behaviour in infants. *Genet Psychol. Monogr.*, **12,** 195 (1932).

FRESH WATER FROM SALINE WATERS
An Engineering Research Problem

By BARNETT F. DODGE
Yale University

IN DISCUSSING this subject I have two main objectives; first, to show the nature of engineering, or applied, research and how it differs from what is variously called basic research, scientific research, fundamental research, or pure research; and second, to present some interesting technical information about a currently very important engineering problem.

There is no clear-cut distinction between applied and pure research. I suppose the main difference lies in the objective. In pure research one is seeking new information to extend the boundaries of knowledge or to develop new correlations to help explain and classify existing knowledge. There is usually no thought of applying the information to a useful purpose although I suppose there is almost always the hope that it will some day be useful to mankind. Engineering or applied research starts with the definite objective of developing something useful for a specific purpose. It may be new data for use in design, a new or improved device or process, or a new product. Perhaps another way to point up this difference is to say that basic research leads to discoveries that are generally not patentable, whereas applied research often leads to patentable results.

The objectives may be the only substantial difference in these two kinds of research. The methods used are quite similar and I believe that engineering research requires the same sort of imagination, creativeness resourcefulness, and initiative as does pure scientific research. I note that many research projects in physics and chemistry are hardly distinguishable from those in some branches of engineering. In both cases it is fundamental data that are being sought and the only difference seems to be the reason for which the data are wanted.

THE SALINE WATER PROBLEM

This is an interesting and vitally important engineering problem that will also serve nicely to illustrate my first theme. Pure water is one of the most important of our natural resources and the necessity of insuring the future supplies needed for our economy can hardly be overestimated. It is both a national and an international problem. It does not have the glamor associated with research on rockets, missiles, and space flight but, in the long run, may be of greater importance to us as individuals.

Basic Facts about Water

Before embarking on a technical discussion of various processes, let us examine a few basic facts about water supplies. The total water available as rainfall on land areas is much more than ample to supply all needs. In fact, it is estimated to be about 30,000 gallons per day (gpd) for every inhabitant of the earth. The trouble is that it is not uniformly distributed. There are many inhabited areas where serious water shortages have existed for many years; there are even some localities that have never before had a water problem and are now beginning to feel the pinch. This is true in some parts of the United States.

The establishment by Congress in 1952 of the Office of Saline Water (OSW) of the Department of the Interior is evidence of the recognition in this country of the importance of this problem. The OSW is concerned with fostering the development of economical methods for the recovery of fresh from saline waters of various kinds. It carries out its task by research contracts with institutions and industrial concerns. It is also committed to a program of construction of five demonstration plants, three for sea water and two for brackish water, with capacities in the range of 100,000 to 1,000,000 gpd of fresh water. These plants will use five different processes and will be located in widely separated areas. All five processes have now been selected and all but one of the sites.* The five processes are: *1*) multiple-effect evaporation in long-tube-vertical (LTV), falling-film evaporators; *2*) multistage flash evaporation; *3*) electrodialysis; *4*) vapor-compression distillation; and *5*) a freezing process. Each of these will be briefly discussed.

Table 1 lists a few basic facts about water supply and

TABLE 1. BASIC FACTS ABOUT WATER

Average sea water contains about 35,000 parts per million (ppm) of dissolved solids
Average brackish waters contain 3,000–5,000 ppm of salts (standard for potable water is 500 ppm of salts)
Average rainfall on land areas = 3×10^{16} gal per year (equivalent to 30,000 gal per day for each inhabitant of the earth)
Water usage per person in the United States:
   750 gpd for irrigation—45% of total
   750 gpd for industrial use—45%
   150 gpd for household use—10%
Large-scale transportation of water costs 5–15 cents per 1,000 gal for each 100 miles
Some water requirements:
   275 tons to make 1 ton of steel
   20 barrels to refine 1 barrel of petroleum
   20 tons to make 1 ton of sulfuric acid

* The fifth site has been selected since this was written.

usage. Sea water varies in composition and total solids to some extent but the figure in the table is considered to be the normal concentration. Brackish waters with too high a salt content for many uses are available from wells in many parts of United States, especially west of the Mississippi. Note that the cost of transporting water by pipeline is such that only those places that are relatively near the sea coast—say not over 500 miles—will be able to afford fresh water from the sea.

General Remarks on Methods of Recovering Fresh Water from Saline and their Cost

There is no dearth of methods. Many have been known for a long time, at least in principle, and are easily practiced on a small scale. The real problem is in scaling-up to large plants and reducing the costs. This is a difficult aspect of engineering research and development where economics must always be the guide [1]. No processes have yet emerged which demonstrate in large plants that they can produce water for less than $1.00 per 1,000 gallons, a figure which is still much higher than the present average costs of municipal water supplies. Many sea-water conversion plants have been built; the majority are relatively small units for shipboard use although there are a few large land-based plants located in areas of extreme water shortage where a high price can be justified.

As a basis for comparison of costs for water by processes to be discussed later, the following average cost figures are useful to have in mind: [2]

Water for irrigation	1 cent per 1,000 gal or $3.24 per acre-foot
Industrial water	2.5 cents per 1,000 gal
Household water	30 cents per 1,000 gal

Actual large-scale plants now in operation producing fresh water from the ocean have costs in the range of $1.50 to

$3.00 per 1,000 gallons. Estimates made by various groups working on new processes for plants of various sizes run from 25 cents to $6.00 per 1,000 gallons. The most optimistic estimate that I am willing to make is that fresh water can probably be produced from sea water at a figure within the range of 50–75 cents per 1,000 gallons in large units, i.e. not less than 10^6 gpd and preferably larger. It is well to note that even at $1.00 per 1,000 gallons, water is a pretty cheap commodity—only 25 cents for a ton. Furthermore, most people can afford to pay considerably more for water than they now do, if it is necessary.

There are two main items in the cost of producing fresh water from saline water by most processes—fixed costs on the investment and energy cost. (In the case of the membrane processes, the cost of membrane replacement is another important item.) No one cost can be considered negligible in any process and they are closely interrelated. It is a fact well known to engineers but frequently overlooked by others that, as one tries to decrease the energy requirements for a process, the fixed costs on the investment are almost inevitably increased, and one of the most important tasks of the process engineer engaged in research and development is to arrive at the optimum balance between these two opposing tendencies. This will be illustrated later in the discussion.

Energy Requirements

Since energy is such an important cost item, it is desirable to consider what is the minimum requirement and how efficient are actual processes. The minimum requirement for a completely reversible process is given by the thermodynamic relation

$$-W_{\min} = \Delta B = \Delta H - T_0 \Delta S \quad (1)$$

where W is work, B is the thermodynamic function known as "availability," H is enthalpy, S is entropy, T_0 is the absolute temperature of the environment or any suitable "heat sink," and Δ indicates a finite difference.

The work calculated from this equation is entirely independent of the nature of the process or the mechanism involved and, furthermore, it does not assume an isothermal process but depends only on the initial state and the final state achieved. For calculation of the work (1) can be put in more convenient form (strictly true only for an isothermal process)

$$W_{\min} = \int_{n_1}^{n_2} RT \ln \frac{p}{p_0} \, dn \qquad (2)$$

where p and p_0 are vapor pressures of salt solution and pure water, respectively, R is the gas constant, and n is the number

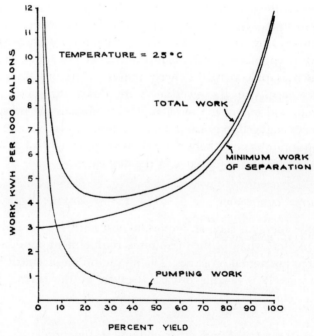

FIG. 62. Minimum work of separation and pumping work as a function of per cent yield of fresh water from sea water.

of moles of water removed. For the special case of zero recovery (i.e. the separation of a very small amount of water from a large amount of saline water) equation (2) reduces to

$$-W_{min} = RT \ln \frac{p_0}{p} \text{ (per mole)} \quad (3)$$

for $t = 25°C$, and a salt concentration of 35,000 ppm NaCl, $-W_{min} = 2.98$ kilowatt hours (kwh) per 1,000 gallons of fresh water.

This is the figure usually cited as the minimum possible work, and the thermodynamic efficiency of an actual process is often based on it. Actual processes will have much larger energy requirements for two main reasons. The first is that a zero per cent recovery is wholly impractical as it would require infinite work to pump the infinite supply of salt water. With the aid of equation (2) one can calculate the minimum or reversible work for various percentage yields of pure water. The results are plotted in Figure 62. On the same graph is plotted the work for pumping the supply of feed water based on an assumed 50 foot head and 70 per cent pump efficiency, and also the total work which is the sum of the two. The minimum separation work increases sharply with increased recovery, and pumping work generally decreases, resulting in a minimum total work at about 40 per cent. Figure 62 is still based on a completely reversible process regardless of per cent recovery. If one uses equation (3) to calculate the work as a function of per cent recovery, higher values will be obtained for all cases except that of zero recovery, because the equation represents the ideal work of separation for a single-stage process which is not completely reversible.

For brackish waters the minimum work requirement is considerably less. For example, for 50 per cent recovery from a feed of 5,000 ppm salt, the minimum reversible work is 0.71 kwh per 1,000 gallons, compared with 4.15 kwh for sea-water feed.

The second reason that actual processes require much more work than the minimum calculated above is that the minimum is based on a *reversible* process—i.e. one that operates with zero driving forces. This is an ideal, limiting case that can be imagined but never realized in practice. The driving forces, which must have finite values for any real process, are such things as temperature differences across heat exchangers, pressure differences for fluid flow, concentration differences, electromotive force (emf) differences, and the like. Whenever a process takes place with a finite driving force, it is an irreversible process and will inevitably require more work than a reversible one. For example, even relatively small temperature differences in a heat exchanger can increase the work very materially. As a result of the cumulative effect of several irreversible effects, all practical processes have a relatively low thermodynamic efficiency—of the order of 2 to 5 per cent based on the minimum work for zero recovery or 3 to 7 per cent based on the minimum work for 50 per cent recovery, with no allowance for feed-water pumping.

It might appear that efficiencies of the order of 2 to 7 per cent leave plenty of room for improvement and that it should be relatively easy materially to decrease the energy requirement of saline-water conversion processes. Such is not the case. Whereas some improvements are being made and will continue to be made, the best that is likely to be achieved in the foreseeable future is an efficiency of about 10 per cent, or a work requirement of about 30 kwh per 1,000 gallons with a sea-water feed. A 20 per cent efficiency, or 15 kwh per 1,000 gallons, is a possible goal but probably unattainable at the present time.

To see the reason for this we have to consider both the engineering and the economics involved. Any decrease in the driving force of an operation such as heat transfer, gas compression, electrodialysis, or the like, inevitably results in an increase in the size and hence the cost of equipment. This is easy to see in the case of a heat exchanger. If the mean temperature difference for heat transfer is halved, the required surface area will be twice as great, other factors remaining the same. In the cost of the recovered water, the two items of energy cost and fixed charges on the investment are the dominating ones in a large plant, and are of about equal importance. If the energy cost is decreased it is, in general, at the expense of an increase in fixed charges and one arrives through an economic balance at the minimum cost for the optimum case. Present indications are that, with the best engineering we can devise, the minimum cost will occur within the range of 10 to 20 per cent thermodynamic efficiency, and probably nearer 10 per cent.

Before leaving the subject of energy requirement it is desirable to point out that the thermodynamically calculated requirements are for energy in the form of work. Whereas some processes do use work directly, most of the evaporation processes utilize energy in the form of heat, and we want to be able to calculate a minimum heat requirement. Minimum work is related to minimum heat through the second law of thermodynamics expressed in the familiar form of the efficiency of a Carnot heat engine, namely

$$W_{min} = Q_{min} \frac{T - T_0}{T} \tag{4}$$

where Q is heat taken into the engine at absolute temperature T, and T_0 is again the temperature of the heat sink.

This shows that the usefulness or availability of heat to do work depends on its temperature level. The calculated minimum heat requirements corresponding to the minimum work requirement of 2.98 kwh per 1,000 gallons for $t_0 = 70°F$, or 1.20 Btu per pound, are shown in the Table 2, along with the pounds of water evaporated per pound of steam at the given temperature.

A single-effect evaporator could evaporate only 1 pound of water per pound of steam used and hence, with steam at

250°F, it would have only 0.5 per cent efficiency and the steam cost alone would be $3.00 to $4.00 per 1,000 gallons. Clearly, some method of reuse of the steam is necessary for a

TABLE 2. MINIMUM HEAT REQUIREMENTS FOR SEPARATING WATER FROM SEA WATER

Temperature at which Steam is Available, °F	Q_{min} Btu/lb	Lb Water Evaporated/Lb Steam
120	13.9	73
150	9.1	109
175	7.3	138
200	6.1	164
250	4.7	211

practical process and several methods will be discussed later. Further details on work and heat requirements of saline water conversion process are given in a paper by Dodge and Eshaya [3].

Energy Sources

Since the cost of energy is such an important item in the cost of fresh water from saline waters, it may be desirable to consider briefly the principal sources of the energy that might be utilized for saline-water demineralization. These are listed in Table 3.

The first, combustion of fuels, requires no discussion as it is by far the most important of our present sources; the second, falling water, can be dismissed with the remark that, if plenty of falling water is available, there is probably not too much need to resort to saline water. Number 3, waste heat, is a large potential source which is untapped, but for much of it the thermal head is too low to allow efficient utilization. To give just one illustration it might be mentioned that the large Lago refinery of the Standard Oil Company of New

Jersey, on Aruba, collects the sea water used as cooling water from various parts of the refinery and subjects it to a flash evaporation at a good vacuum and recovers 5 to 6 per cent of it as fresh water.

TABLE 3. POTENTIAL ENERGY SOURCES FOR SALINE WATER DEMINERALIZATION

1. Combustion of fuels
2. Falling water
3. Waste heat from industrial processes and power plants
4. Nuclear fission
5. Exhaust steam (by-product power from turbine)
6. Solar
7. Waves and tides
8. Wind
9. Geothermal
10. Thermal energy of ocean
11. Fuel cells
12. Concentration difference of salt solutions

Nuclear fission used in specially designed nuclear reactors to generate low-pressure steam has been investigated from an engineering and economic viewpoint and a low cost of 42 cents per 1,000 gallons predicted for a capacity of 50×10^6 gpd of fresh water from sea water [4].

In localities where both electric energy and fresh water are in short supply, it is advantageous to generate high-pressure steam either with nuclear or conventional fuels and use the steam to generate power in a turbine, exhausting at a pressure high enough to permit utilization of the exhaust steam in a multiple-effect evaporator for producing fresh water. Either the water or the electric energy might be considered the by-product. An operating plant using such a process for water and by-product power will be mentioned later.

Solar energy is sometimes considered to be "free" but it is actually more expensive in most applications than either fossil fuels or nuclear fuels when delivered in a usable form.

This is due to its intermittent nature and its low concentration which call for large and expensive collecting surfaces. A considerable amount of work on the use of solar energy is being done in various parts of the world but its use for saline-water demineralization will probably be limited to special applications and special areas where fuels are not available at reasonable cost. For further information on the use of solar energy for demineralization of saline waters, several references are available [5-8].

Tide power, wind power, and geothermal energy from hot layers within the earth with access to the surface, are all possible sources but with very limited, if any, application within, say, the next ten years.

Thermal energy stored in the oceans is a vast supply but most of it is unavailable for lack of a suitable temperature difference. In the tropics it is a well known fact that there is a temperature difference as much as 40°F between the surface water and water at a considerable depth, say 1,000 to 2,000 feet. If this deep water could be brought to the surface without serious increase in its temperature, power could be generated by the steam evaporated from the warm water and condensed by the cold water, and the condensed steam would be a by-product of fresh water. Such a scheme was proposed many years ago by the French engineer Georges Claude and tried out by him in 1926 off the coast of Cuba. A considerable sum of money was spent but the plant was plagued by mechanical difficulties and in the end was a complete failure. In recent years the French government* has revived the scheme and an experimental plant was built at Abidjan on the west coast of Africa. Published results on it are meager and we are not sure if this project is still under way. The idea is an imaginative and intriguing one but, in our opinion, not likely to achieve much practical success.

* This project has been under the direction of Energie des Mers, a partly government-owned organization.

Fuel cells have been much in the news lately but they have a long way to go before they can find economical application on a large scale. Essentially they are nothing but electrochemical cells using carbonaceous compounds or hydrogen as the fuel instead of a material like zinc which is used in ordinary dry cells. They can turn the chemical energy of the fuel directly into electrical energy, thus by-passing the boiler, steam turbine, and generator of the conventional power plant and also achieving a higher efficiency (75 to 80 instead of 30 to 40 per cent), since the limitation imposed by the second law of thermodynamics is removed because the energy is not converted to a thermal form. Eventually, fuel-cell plants may be competitive with fuel-using power plants of the present type but this is probably a long way in the future.

Source 12 probably has very little potential for practical application but it is interesting and deserves passing mention. The basis for it is the free-energy decrease, ΔF, when a salt solution is diluted in a reversible manner, which means in an electrochemical cell so that an external emf is produced. The basic relation is

$$W = -\Delta F = RT \ln \frac{a_2}{a_1} \tag{5}$$

where a_2 is the activity of salt in the concentrated solution and a_1 that in the diluted solution. The use of the method for desalination of sea water presupposes the existence of a brine more concentrated than sea water. This could be brine from a salt mine, or produced by solar evaporation of sea water, or perhaps the concentrated brine produced in another saline-water desalting process. The use of this concentration driving force is the basis for the "osmionic process" to be discussed later. In this process the concentration difference is used directly without the production of an external emf.

Processes in Actual Use or Proposed

General. Many processes have been proposed (a report from the Office of Saline Water in 1952 [9] listed 32 if we include phenomena that might be the basis for processes) but only a few are in actual use and probably only a very few of the many potential processes will ever reach the stage of practical application. One of the more fantastic proposals is to tow icebergs from the Arctic regions and melt them in dry docks at a point in the temperate zone. This has been seriously proposed but no engineering and economic analysis has been

TABLE 4. ACTUAL AND POTENTIAL PROCESSES FOR SEPARATION OF WATER FROM SALINE WATERS

I. Processes that separate water from a brine
   A. Evaporation or distillation. Many variants depending on conditions, type of equipment, and source of energy.
      1. Single-stage flash evaporation
      2. Multistage flash evaporation
      3. Multiple-effect submerged tube
      4. Vapor compression
         (*a*) Submerged tube
         (*b*) Flash evaporation
      5. Combination of vapor-compression and multiple-effect evaporation
      6. Solar energy
      7. Nuclear energy
      8. Ocean temperature difference
      9. Critical pressure
   B. Freezing or crystallization processes
      1. Indirect refrigeration
      2. Direct refrigeration using a special refrigerant
      3. Direct refrigeration using water as refrigerant
      4. Hydrocarbon hydrates
      5. Zone freezing
   C. Solvent extraction (liquid-liquid interface)
   D. Reverse osmosis (semipermeable membranes)
II. Processes that separate salt from a brine
   A. Ion exchange
   B. Electrodialysis with perm-selective membranes
   C. Osmionic process

made as far as I am aware. Common sense seems to indicate that it would not be a competitive method.

It is convenient to classify all processes into two categories: *1*) processes that separate water from the brine, and *2*) processes that separate the salt. Since the salt is present in small concentration, the processes of category *2* appear to have a theoretical advantage, but this does not work out in practice. It is also useful to think of all processes as involving some sort of a barrier that is permeable to water but not salt, or vice versa. For example, in all distillation processes a liquid-vapor interface is an effective "semipermeable membrane," allowing water molecules to pass freely in either direction but barring the passage of salt molecules. The same role is played by a liquid-solid interface in all of the crystallization processes. In processes such as reverse osmosis and dialysis, actual physical membranes provide the barrier. In Table 4 there are listed a number of the more important processes that are either in actual use or have been seriously considered for use [9].

In addition to these processes there are given in Table 5 [9] a number of phenomena which have been considered as offering possible bases for future processes. No one of these

TABLE 5. PHENOMENA THAT MIGHT PROVIDE A BASIS FOR A PROCESS OF DESALTING

Sublimation of salt
Adsorption of salt on solids
Thermal diffusion
Ultrasonic vibrations
Underwater combustion (with a distillation process)
Ion repulsion at water-oil boundary
Selective solvent for salt
Electrolysis
Streaming potential
Electrostatic effects
Electromagnetic effects
Ultrahigh-frequency current

latter has yet been developed even to the point of a laboratory-scale process, and they are listed merely to indicate the scope of the search that is under way for new processes. None of these appears very promising at the moment, but a new discovery might put one of them into the running.

It will not be possible to discuss any of the processes listed in Table 4 in detail but a few high lights on some of the more important or more interesting ones are in order.

Distillation or evaporation processes. These are the most important and most extensively used at the present time. Distillation is also the method by which nature produces fresh water from the ocean, using solar energy. Thousands of relatively small units are in use on board ship and some of these are as large as 50,000 gpd. Similar land-based units are in operation at various locations around the globe. In all distillation processes heat must be supplied to effect the separation and this is usually in the form of steam which condenses and gives up its latent heat. The water can either be directly evaporated at constant temperature by submerging heating tubes in the boiling liquid, or energy can be stored in the liquid by allowing its temperature to rise while the pressure is maintained high enough to prevent boiling. Then, upon suddenly lowering the pressure, a small fraction of the liquid is suddenly vaporized. This is known as "flash evaporation."

If distillation is carried out in only one stage, not more than one pound of evaporated water is obtained per pound of steam used, and the cost for energy alone would be prohibitive as has already been shown. For a distillation process to be economically feasible the steam must be reused. It is easy to see in principle how this can be done. If steam condensing at 220°F is boiling a salt solution at 210°F (a finite temperature difference is necessary for a reasonable rate of heat transfer), substantially all the energy contained in the steam is still present in the water vapor produced, and the only difference is that it is at a lower temperature and no thermal head exists to enable it to evaporate more water at 210°F. It could, however, evaporate more water from a solution boiling at, say, 200°F. Thus, by providing a series of evaporator bodies (usually called "effects"), each one at a lower pressure than the preceding one and therefore with decreasing boiling temperatures, it is possible to reuse the steam a number of times and greatly improve the economy. Such a scheme is known as multiple-effect evaporation and a 4-effect system is shown diagrammatically in Figure 63.

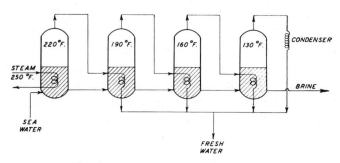

Fig. 63. Multiple-effect evaporation [2].

Theoretically one should get n pounds of water evaporated per pound of steam in an n-effect system but, due to various losses, the actual pounds evaporated, W, are more nearly equal to

$$\frac{W_1(1 - W_1^n)}{1 - W_1}$$

where W_1 = pounds of water evaporated per pound of steam in a single effect. W_1 is generally in the range of 0.85 to 0.95; with $W_1 = 0.93$ and $n = 15$, $W = 8.8$ pounds water per pound of steam. Assuming $W_1 = 0.95$, it is easy to calculate that the maximum possible yield is $W = 19$.

There is a theoretical as well as a practical limit to the number of effects or stages. Because of the boiling-point elevation (BPE) of a salt solution, and the fact that this BPE is lost as far as being useful for heat transfer is concerned, there is a definite upper limit to the possible number of stages. Thus we can write the relation

$$n = \frac{t_1 - t_2}{(\text{BPE})_{\text{av}}} \qquad (6)$$

where t_1 = temperature of steam to the first effect and t_2 = temperature of the heat sink. For normal sea water concen, trated twofold and for a temperature range of 80° to 212°F. $(\text{BPE})_{\text{av}}$ is about 1.27°F and $n = (212 - 80)/1.27 = 104$. Economics, however, fixes a much lower limit to the number of effects. The largest number now in use in any sea-water distillation plant is that in the 6-effect system on the island of Aruba in the Netherlands Antilles with a capacity of nearly three million gpd of fresh water produced in three units. The writer visited this plant in November 1958 while it was being erected and took the pictures shown in Figures 64 and 65. It may be of interest to note that the Netherlands Antilles

FIG. 64. Plant under construction on Aruba.

government has recently issued a series of commemorative stamps with a picture almost identical with that in Figure 65.

An interesting and important feature of the Aruba plant is worth mentioning. The island has a shortage of electricity as well as water and instead of generating low-pressure steam (20 to 25 pounds per square inch) which is the heating medium for the distilling plant, steam is generated at high pressure (850 psi) and first used in a turbine to generate power. The turbine exhausts the steam at 20 psig with about 80 per cent of its heat content still present. In this way about 110 kwh are generated as a by-product per 1,000 gallons of fresh water. Assuming a net worth of 0.5 cent per kwh for the electric energy, it is easy to see that this by-product gives a substantial reduction in the cost of water. Even with the by-product power, the thermodynamic efficiency of this process based on the reversible work for 40 per cent recovery of fresh water is only 2.3 per cent [3] and this is one of the most modern of the large-scale units now in operation.

Fig. 65. Closeup view of evaporators in Aruba plant.

Recent engineering calculations [2] indicate that the optimum number of effects lies between 10 and 20. The first demonstration plant of the OSW to be located on the Gulf coast is to be of the falling-film, multiple-effect evaporation type.* Badger and Standiford [10] have designed a 12-effect LTV evaporator which would be operated in connection with a 60,000 kw power plant. With power at 5.13 mills and steam at 12.2 cents per 10^6 Btu, this plant is estimated to produce 17 million gallons of water per day at the unusually low figure of 23 cents per 1,000 gallons. How closely this figure can be approached in practice remains to be seen.

In flash evaporation the same general principles apply and more than one stage is essential for good energy economy. The economical number of stages in flash evaporation is considerably greater than for ordinary multiple-effect boiling evaporation with the same over-all Δt. For example, one optimum case was calculated for 52 stages. Although flash evaporation is presumably less efficient thermodynamically than boiling-type evaporation, it is more flexible in that the performance is not definitely fixed by the number of stages.

Any source of water—water from power plants, chemical plants, or Diesel-engine cooling water—can be flash evaporated to produce fresh water. With temperature differences of 60°F the yield is low, about 6 per cent. If water production is to be combined with power production, as in the case of the French plant at Abidjan, the yield is considerably less because the flashed vapor must have a pressure higher than the final pressure in the system if any power is to be generated.

In a single stage, the flashing occurs at the lowest possible temperature and no recovery of the latent heat is possible since no thermal head exists. By flashing in several stages, each at a lower pressure and hence lower temperature than the preceding one, the latent heat of each increment of

* This plant has now been constructed and has recently been put into operation.

flashed vapor can be utilized to preheat the feed stream, and live steam is used only to give the final boost to the sea-water feed before it enters the first flash stage.

A considerable number of multistage flash units is either in operation or in the construction or planning stage. The largest in actual operation is the 2.5 million gpd, 4-stage unit in the sheikdom of Kuwait on the Persian Gulf. This type of evaporation appears to offer a number of advantages, two of the most important being the relative ease of "scaling-up" to very large units, and the minimization of the scale-formation problem as a result of the fact that the evaporation does not occur at the heating surface. Engineering design for cost-estimating purposes has been carried out on a very large plant [4] with 52 flash stages, a capacity of 50 million gpd, and a nuclear reactor furnishing the steam. The estimated cost of water was 42 cents per 1,000 gallons.

Vapor-compression evaporation. There is another well known method of recovering the latent heat of the steam and that is by the use of "vapor-compression" or "thermocompression" evaporation. The principle is very simple and is illustrated in Figure 66. As pointed out before, the over-all process in an evaporator involves nothing but a throttling expansion of steam at constant enthalpy. The latent heat can be recovered if the pressure of the steam is returned to its original value and this is easily done by means of a compressor. As shown in the diagram, the water vaporized from the brine solution is compressed, desuperheated and sent to the tubes of the evaporator where it condenses and gives up its latent heat. The diagram shows a submerged-tube type of evaporator but it can also be used with flash evaporation. The advantage of this system is that only one evaporator is necessary for good energy economy as compared with six or more in the multiple-effect system but, of course, this is partly offset by the need for a compressor and by the requirement of mechanical rather than thermal energy input.

Since the energy for compression is an important cost item, the pressure difference across the compressor must be kept as low as possible. Since the pressure difference is directly

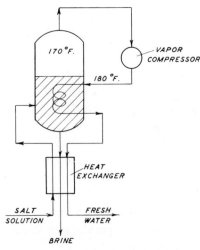

FIG. 66. Principle of vapor-compression evaporation [2].

Vapor compression evaporator

related to the temperature difference across the heating surface of the evaporator which, in turn, controls the capacity, this is another place where economic balance is used to determine the optimum operating conditions. It turns out that the temperature difference should be of the order of 5–10°F and this corresponds to a pressure ratio of 1.15 to 1.25. Since, for the prevention of scale, it is desirable to operate at a temperature well below the normal boiling point, the compressor will be required to compress very large volumes against a small head and the high-speed rotary axial or centrifugal compressors are ideally suited for this type of service.

It is of interest to note that the maximum possible economy for a multiple-effect evaporator (i.e. with an infinite number of effects) is about equivalent to a vapor-compression system with a 15°F temperature difference. This means that the energy cost for the vapor-compression evaporator will be less than that for any multiple-effect system if the temperature difference in the former is less than 15°F. This comparison, of course, depends on a number of factors and assumptions, the most important of which are the unit costs of steam and electric energy, for which we have used the figures of 55 cents per 1,000 pounds and 7 mills per kwh, respectively [3].

A disadvantage of the vapor-compression process as compared with multistage evaporation is the need for electric energy or for high-pressure steam to operate a turbine drive, and the need for a source of steam to start the operation. Proposals have been made to combine vapor-compression and multistage evaporation by generating high-pressure steam for use in a turbine exhausting at a somewhat elevated pressure. The power generated is used to operate a vapor-compression evaporator and the exhaust steam operates the multieffect evaporator. This combination is advantageous where no other use exists for the power that can be developed from the high-pressure steam.

Solar evaporation. A fresh-water plant with a capacity of 6,000 gpd using the radiation from the sun was operated at Las Salinas, Chile, as far back as 1872. Recently there has been a great revival of interest in use of solar energy, and a considerable research and development effort is going on around the world. Only a brief discussion will be given here because it is not expected that solar energy can be competitive with other energy sources for the large-scale plants with which this paper is mainly concerned.

Although solar energy is "free" in one sense, it is actually a relatively expensive form of energy when it is collected and put into useful form. It has two very serious limitations: its low concentration and its intermittent nature. To produce 10^6 gpd of fresh water at about 45° latitude, would require something of the order of 10^7 square feet of collecting surface based on 0.8 pound water per day per square foot.

At a cost of $1.00 per square foot which may be somewhat optimistic at present, the investment in collecting surface alone would be $10,000,000. This may be compared with an estimated total investment of about $2,000,000 in a vapor-compression plant of the same capacity using fuel oil as the energy source [11]. Other types of distillation plants may involve capital investments of no more than $1,000,000. A simple solar still is illustrated in Figure 67. Many variations

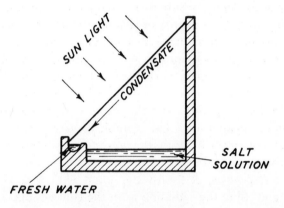

FIG. 67. Simple solar still [2].

of this simple design have been made to increase the effectiveness, but more complicated designs increase the investment and our old friend economic balance is again at work.

To summarize the situation with regard to solar distillation, it seems reasonable to state that this method will find application in relatively small units in places where fuels are scarce and expensive. It offers little promise for large plants in areas where coal, oil, or gas is available, or where nuclear reactors could be built.

Critical-pressure process. This is another rather impractical process, but the principles involved are so interesting that it is worth more than passing mention. In the ordinary evaporation process, most of the heat required (i.e. the latent heat) is supplied at constant temperature and hence it cannot be recovered by simple heat exchange which requires a temperature difference. If the sea water is pumped through a heat exchanger at a supercritical pressure (i.e. above 3,200 psi) no vaporization occurs and heat is added with a continually rising temperature. If heated to a temperature above

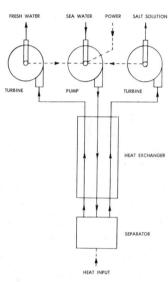

FIG. 68. Schematic diagram of supercritical-pressure process Center unit at top is a pump, outer units are turbines (p. 81 [6]).

the critical temperature of water (705°F) a phase separation occurs without any appreciable latent-heat effect. A simplified schematic diagram is shown in Figure 68. It will be noted that some heat is added at the phase separator. This is obviously necessary in order to have a temperature difference in the exchanger but it is a relatively small fraction of the total enthalpy change. Data on the phase compositions for a synthetic sea water as a function of pressure and temperature have been published [12]. One isolated set of values may be of interest. At 750°F and 3,400 psia, the lighter phase contained 575 ppm of salt and the denser one 260,000 ppm, or 26 per cent. It is evident that a good separation can be made. Because of the high pressure with the attendant high cost of pumping, the use of turbines for power recovery is an essential part of the process. This process might have promise except for two very serious problems—namely, severe corrosion and very rapid scaling of the transfer surfaces. Both of these appear insoluble (i.e. with reasonable economics) at the present time and research on this process has been abandoned.

Crystallization processes. Processes in which the separation barrier is a solid-liquid interface, permeable to water and not to salt, are of two general types. The first consists of the freezing processes in which the solid phase is pure water (as long as the temperature remains above the eutectic temperature) produced by extracting the latent heat of freezing. The second type is a process in which the solid phase is some compound formed between water and another material, such as a hydrocarbon forming a hydrate.

Three variants of the freezing process are listed in Table 4. The first involves a conventional refrigeration cycle with the refrigerating fluid in indirect, heat-exchange relationship with the brine and ice systems. The latent heat extracted from the freezing brine by the evaporating refrigerant is "pumped" up to ice temperature and discharged to the melting ice to form the pure water product. It is clear that the heat removed in the freezing should be discharged at the lowest possible temperature in the interest of economy of energy, and that the ice is a heat sink readily available, since the ice must be melted to produce the desired product. An over-all energy balance will show that not all of the heat can be discharged to the ice. That which corresponds to the work of compression, to heat leaking in from the surroundings, and to temperature differences at the warm ends of exchangers must be discharged to cooling water. So we have two heat sinks in a freezing process, the ice itself, and cooling water or the atmosphere.

An interesting variation of the freezing process is the use

of direct evaporation and condensation of the refrigerant in contact with the brine and ice, respectively, without the interposition of a heat-exchange surface. This is expected to simplify the equipment and also permit the use of smaller temperature differences between evaporating refrigerant and brine, and between condensing refrigerant and ice. It goes without saying that the refrigerant must be very insoluble in water if serious losses of it are to be avoided. A process developed at Cornell University under contract with the Office of Saline Water, using isobutane as the refrigerant is diagrammed in Figure 69. In the ice generator or freezer, liquid isobutane is vaporized from a spray of droplets dispersed in the sea water. Ice produced from sea water in any of the freezing processes is in the form of very fine crystals and, whereas they are free from salt, mother liquor to the extent of 25 to 65 per cent by weight of the ice will cling to the crystals and must be removed by washing them with some of the product. This may be accomplished in a countercurrent wash tower as shown in Figure 69. The isobutane vapor leaving the generator goes to a compressor after which it is partially condensed by the ice which has accumulated at the top of the wash column. About 15 per cent of the isobutane remains uncondensed at ice temperature and it is further compressed and condensed by sea-water cooling.

Disadvantages of this process are the solubility of the refrigerant in water, which is large enough so that some loss will occur, the necessity of using some of the product water to purify the ice, the possibility of the formation of hydrates that may plug passages, and the fact that it is difficult to produce very pure water, as is easily accomplished by a vaporization process. The effect of these and other factors governing the operation of the process will have to be investigated on a plant scale before the feasibility of this process can be established. The outstanding advantages are the elimination, or at least amelioration, of the scale problem and the lessening of corrosion. This process is at present being transferred from the laboratory to the pilot-plant stage.

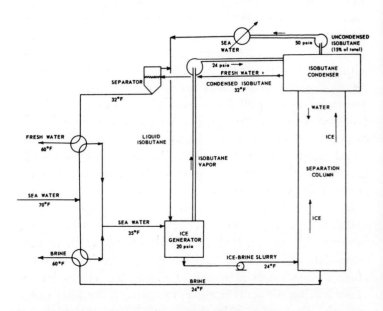

FIG. 69. Refrigeration process using direct contact of a refrigerant immiscible with water (p. 379 [6]).

It is often stated in the literature that the freezing process has a potentially higher energy efficiency than an evaporation process, because the latent heat of freezing is only one-seventh that of the latent heat of vaporization. This is an erroneous statement based on a misconception of the thermodynamics of the process. As has been pointed out in connection with the vaporization processes, the energy requirement is not directly related to the latent heat. In fact both of these processes are essentially heat pumping, and the work required for heat pumping depends not only on the quantity of heat to be pumped but also on the temperature interval over which it is pumped. The fundamental relation for heat pumping based on the first and second laws of thermodynamics is

$$W_1 = Q \frac{T_1 - T_2}{T_2} \tag{7}$$

where W_1 is the work required to pump heat Q from absolute temperature T_2 to T_1. In the case of the refrigeration process this work itself appears as heat at T_1 and it must be pumped to sea-water temperature, T_0. The relation for the total work W_2 in this case is

$$W_2 = Q \left(\frac{T_0}{T_1}\right)\left(\frac{T_1 - T_2}{T_2}\right) \tag{8}$$

Considering an idealized vaporization process with a minimum of irreversible effects, $(T_1 - T_2)$ is the BPE which is about 1.6°F at the 212°F boiling point level for a 7 per cent brine with 50 per cent water recovery. In the idealized freezing process, $(T_1 - T_2)$ is about 8°F. Using these figures along with the other appropriate data, one can calculate that $W_1 = 5.8$ kwh per 1,000 gallons and $W_2 = 6.3$. There is a small difference in favor of the vaporization process but this difference is insignificant in view of the much greater amounts of work needed to compensate for the irreversibilities. Making reasonable allowances for the irreversible effects or driving forces present in any practical process, one would conclude from calculations that the freezing processes require more energy, or less energy, than a vaporization process, depending on the assumptions. A realistic calculation of this sort indicates that a freezing process will require about 38 kwh per 1,000 gallons instead of the 6.3 given above for an idealized process [3]. Careful design could conceivably reduce this somewhat. From figures given in a report to the Office of Saline Water [13] based on a pilot plant operating a direct, water-vapor compression freezing process, we have estimated about 47 kwh per 1,000 gallons for a plant capacity of 10^7 gallons per day.

This last-mentioned variant of the freezing method dispenses with a refrigerant and uses direct freezing at low pressure. This process has been developed by the Carrier Corporation in this country and by Zarchin in Israel, and it is possible that this process will be selected for the fifth demonstration plant of the OSW. A diagram of one form of the process is shown in Figure 70. In this process, the pre-

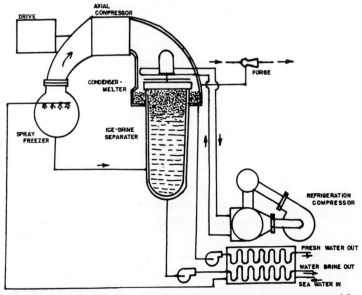

FIG. 70. Direct freezing process using vapor compression (p. 386 [6]).

cooled sea water is sprayed into a vacuum chamber whose pressure is maintained lower than the vapor pressure of the brine solution of the desired final concentration. Some of the water will vaporize and, since the process is adiabatic, the necessary heat of vaporization is taken from the solution itself, cooling it and forming ice. The ice-brine slurry flows to a wash column where the ascending ice crystals are washed by descending pure water produced by condensing the vapor from the compressor which follows the freezer. As in all freezing processes, some of the heat must be discharged at ambient temperature and an auxiliary refrigeration system must be provided to do this.

One of the main disadvantages of this process is the very low pressure required in the freezer, condenser, and wash column. This is of the order of 0.005 atmosphere in the freezer and one can readily calculate that for a plant producing 10^6 gpd the compressor would have to handle about 3 million cubic feet per minute of water vapor. This is well beyond the capacity of any existing compressor units, not to mention the difficult problem of constructing and maintaining vacuum-tight the other bulky items of equipment required. Nevertheless, the process has considerable promise and may well turn out to be competitive with the distillation processes.

TABLE 6. EQUILIBRIUM CONDITIONS FOR FORMATION OF HYDROCARBON HYDRATES [14]

Hydrocarbon	Temperature, °F	Pressure, psia
CH_4	32	383
CH_4	56	1540
C_2H_6	32	76
C_2H_6	56	400
C_3H_8	32	24
C_3H_8	42 (maximum temperature for hydrate formation)	80

Very little work on the hydrate process has been published,* but we understand that some pilot-plant studies are in progress. The equilibrium conditions for hydrocarbon-hydrate formation have been investigated and some figures given in Table 6.

* Since this was written, a paper [W. G. Knox et al., *Chem. Eng. Prog.*, **57,** 66 (1961)] has appeared that gives considerable information on the hydrate process developed by the Koppers Company.

The conditions for formation of hydrates of normal and isobutane are not known with certainty. If they do form, it appears that they are stable only over very narrow ranges of pressure and temperature.

Other compounds forming hydrates, such as halogenated derivatives of methane and ethane, are also being considered. Some of these are known to form at temperatures above 60°F at relatively low pressures (i.e. of the order of 50 to 100 psi). For example, $CH_2Cl\,F$ forms a hydrate whose decomposition temperature is 64.2°F with a corresponding pressure of 41.5 psia.

Presumably the only advantage of such a process over freezing is that higher temperatures can be used and refrigeration avoided, but this is likely to bring in the disadvantage of elevated pressures. Very little is known about the kinetics of the reaction and there are so many unknown factors about this process that it is not possible at present to assess its practical value. My opinion, for what it is worth, is that it will not be competitive.

Zone freezing (analogous to the zone melting of metals which has been so successful in producing ultrapure metals) has been tried on sea water but the results showed that it has little promise [15].

Solvent extraction. This is a method of great importance in the process industries and it is natural that it should be considered for the demineralization of saline waters. It would be logical to extract the salt from the water but no suitable solvent is known and the search seems rather hopeless. However, solvents are known—notably *n*-methyl-*n*-amylamine and *n*-ethyl-*n*-butyl-amine—which will selectively absorb water from a salt solution. Energywise the process does not appear promising because the only reason that the water will leave the salt solution and go into the solvent is that the activity is lower in the latter. Consequently, all we have succeeded in doing is transferring the water from one solution to another one from which the theoretical work required to separate it is greater than from the original one. A rough comparison of energy requirement with that for distillation or freezing may be made as follows: assuming that the heat of solution is about 50 Btu per pound of product water and that this heat is to be pumped over a range of 50°F, use of equation (7) gives a theoretical energy requirement of 11 kwh per 1,000 gallons compared with about 6.0 for the other two processes mentioned. In spite of this, considerable work is being done on the process and some consider it to have promise [16].

Before dismissing the subject, a few figures may be of interest. With an amine solvent, the extract in equilibrium with sea water at 90°F will contain approximately 70 mole per cent water and 30 per cent amine, with less than 1 per cent salt. The waste sea water will contain less than 0.1 per cent amine but, since the latter is an expensive material it will have to be recovered. If the extract is heated to about 140°F it will separate into two phases: a water phase with small amounts of salt and amine and an amine phase with about 60 per cent water.* The amine phase is recycled to the extractor but the water phase must be further treated to recover amine, and its salt content is also too high for many uses. In addition to these disadvantages, the process requires the transfer of very large amounts of heat in cycling between, say, 90° and 140°F, and this will require such a large investment in heat-exchanger surface that this factor alone appears to make the process unattractive. There are some potential advantages associated with this process that may balance, to some extent, the disadvantages. For example, it can be operated with low-grade thermal energy and may give less trouble from scale formation and corrosion than do the distillation processes.

* Since this was written more data on the process have appeared in a paper by Hood and Davidson [*Advances in Chem. Ser.* **27**, 40 (1960)].

Reverse osmosis. It is well known that if a salt solution is separated from pure water by a membrane permeable only to water, water will permeate the membrane and dilute the solution. If the solution volume is held constant, a pressure will develop in the solution and the pressure at equilibrium between water and the solution is known as the osmotic pressure. Its value can be calculated from the principle that the activities of water in the two phases must be equal at equilibrium and from the known effect of pressure on the activity of a salt solution. For sea water at ambient temperature, the osmotic pressure is about 22 atmospheres but this would correspond to zero recovery in a desalting process. For 50 and 75 per cent recoveries, the osmotic pressures are 43 and 84 atmospheres, respectively.

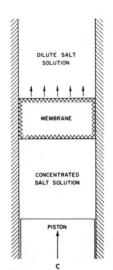

FIG. 71. Principle of the reverse-osmosis process (p. 180 [6]).

The process is reversible and if the pressure on the salt-solution side of the diaphragm is increased above the osmotic pressure by use of a piston, as shown in Figure 71, the flow of water can be reversed and pure water squeezed out of a salt solution. The process requires no heating or cooling but only the application of mechanical energy and is very simple in principle. It is sometimes looked upon as an ultrafiltration process, but it is more than just filtration because of the osmotic effect. To obtain practical rates, the pressure would have to be appreciably higher than the equilibrium or osmotic pressure, and hence, for a 50 per cent recovery, it is likely that a pressure of at least 800 psi would be required. This not only imposes a very severe condition on any thin membrane but also probably makes it necessary, or at least desirable, to use turbines to recover some of the pumping power.

For the case of 50 per cent recovery of pure water, and assuming 70 per cent pump efficiency and a pressure of 800 psi, the calculated work would be 16.6 kwh per 1,000 gallons with no power recovery. Assuming only 50 per cent recovery of power, which is very reasonable, the net energy requirement is only 11 kwh per 1,000 gallons, much the lowest of any process. In spite of this, the process is not practical at the present time nor likely to be for some time to come because of the lack of a suitable membrane. Membranes of cellulose acetate, cellophane, or similar materials will perform the desired separation, but their capacity is very low—far below the capacity of the phase barrier in distillation or freezing— and they deteriorate rapidly in use. From some data in the literature we have estimated that the rate of production with present membranes is of the order of 0.01 gallon per hour per square foot which is less than $\frac{1}{100}$ of the rate that is easily obtained by distillation.*

Research is going on to develop better membranes, but it is a very difficult problem as one can see by considering the fact that the membrane must have very fine pores to hold back the salt, and at the same time pass water at a good rate, which demands relatively large pores. The mechanical problem of supporting a thin membrane to withstand such a large pressure difference is a difficult one that has been fairly well solved. The potentialities of the process are such that continued effort on the development of membranes is well justified.

Processes that separate salt from saline water. These are much less numerous than the processes that remove the water directly. The first of those listed under Section II in Table 4 (ion exchange) is a widely used process for water purification but it is not at present applicable to saline waters, because the cost of chemicals for regeneration of the ion-exchange

* A recent news item (*Chem. Eng. News*, April 11, 1960, p. 64) mentions new membranes that give 20 times this rate.

resins becomes prohibitive when solutions of the concentration of sea water or even brackish waters are treated.

Electrodialysis. This process has been developed rapidly in the past several years and is now the most practical and most widely used one for the treatment of brackish waters. At present it is not very suitable for the treatment of sea water for reasons that will be apparent later.

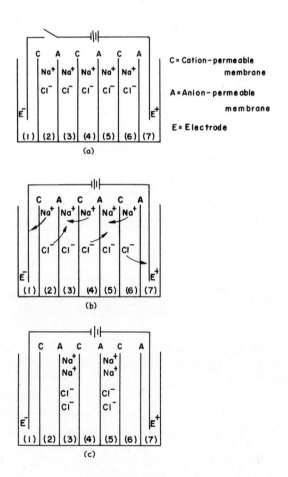

Fig. 72. Principle of the electrodialysis method.

The simple and ingenious principle of this process is illustrated in Figure 72: Consider an assemblage of parallel compartments (1, 2, 3, etc.) separated by thin membranes, which are alternately cation- and anion-permeable. The cation-permeable membranes (C) contain negatively charged ionic groups as in anion-exchange resins, and hence tend to repel the anions but allow cations to permeate through them. Conversely, the anion-permeable membranes (A) have positively charged ionic groups which repel the cations. The initial situation is shown at (a), namely, a uniform concentration of Na^+ and Cl^- ions in all compartments. When an electromotive force is imposed across the assembly by electrodes, E, all Na^+ ions move toward the anode and all Cl^- ions toward the cathode, but the movement is impeded by the membranes in the following manner: Na^+ ions can move out of (2) through C into (1) but those in (3) cannot move out of this compartment because they cannot penetrate an A membrane. Likewise Cl^- ions can move from (2) to (3) but those in (3) are prevented from moving out by the C membrane. The net result as shown in (c) is a depletion of both Na^+ and Cl^- ions in the even-numbered compartments and a concentration in the odd-numbered ones, so that alternate compartments are filled either with a dilute solution (pure water at the limit) or a concentrated brine. (Other ions behave similarly but, for simplicity, the saline water is assumed to contain only Na^+ and Cl^- ions.)

In an actual unit several hundred of these thin membranes, alternately of the A and C types, are separated by spacers of a plastic material and the whole assemblage placed in a press to hold the membranes in a fixed position. Suitable electrodes for introducing the electric current and channels for flow of saline-water feed, fresh water, and concentrated brine are provided. Fluid flow is in parallel through all the compartments as in the case of a filter press.

The heart of this process is the membrane, and considerable research effort has been and is being devoted to its improvement. Desirable properties of membranes for application to large-scale demineralization of saline water include high electrical conductivity, low permeability to the passage of water, high selectivity to either anions or cations, good mechanical strength, good chemical stability to any trace impurities in the water such as H_2S or Fe, and low cost. Some of these requirements are incompatible with each other and a compromise is necessary in arriving at a membrane satisfactory from the over-all point of view. The picture is further complicated by the fact that the desirable characteristics of a membrane are influenced to a large degree by the concentration of the solution being demineralized. This explains the great variety of membranes now being used or under investigation for use in large-scale demineralization units.

The first ion-selective membranes were the naturally occurring ones of plant and animal origin. The collodion membranes prepared by Michaelis (see review paper by Sollner [17]) were probably the first to be synthesized. Sollner and co-workers were successful in preparing films which were not only highly ion-selective but also of high ionic permeability. A great variety of membranes of different characteristic properties has been developed since then. These membranes fall into two general categories, homogeneous and heterogeneous. The first are essentially crosslinked resins and are characterized by high ion concentrations which account for their very high selectivity. Their mechanical properties are, however, generally unsatisfactory, unless the degree of cross linking is decreased, and this results in a loss of selectivity and higher ohmic resistance. The heterogeneous membranes are prepared by incorporating ion-exchange resin particles in an inert plastic matrix. Such membranes can be prepared by continuous and relatively inexpensive processes; however, they also have certain disadvantages: their ohmic resistance is high and relatively thick films are needed to give the necessary mechanical strength. Dimensional instability due to swelling of the resin particles may lead to rupture. A relatively new type known as an interpolymer membrane has been developed by Gregor [18] by casting a film from a homogeneous solution of two polymers, one a polyelectrolyte or an ionogenic polymer, the other a water-insoluble film-forming plastic.

Some examples of membranes with different ionic groups incorporated in them include quaternary ammonium anion-permeable membranes, sulfonated cross-linked polystyrene resins, polythene films treated with chlorosulfonic acid, films containing positively charged sulfonium groups, pyridinium membranes prepared by treating parchmentized paper with a suitable pyridinium compound, and so forth. The concentration of functional ionic groups in the membranes depends largely on the type of membrane and the mode of preparation, and it may vary anywhere from one to seven molal.

The calculation of the energy requirement for this process is complicated by the many factors affecting it. The total energy requirement is made up of the electrical energy and the mechanical energy required to move the fluids. Both of these are affected by the nature of the membrane, the number of membranes in a stack, the spacing, the fluid velocity past the membranes, the salt concentration, and the current density. It would lead us too far afield to enter into a quantitative discussion and for further details the reader is referred to several papers [19–22]. A brief qualitative discussion may be of some interest in again showing how the principle of economic balance controls. At the capacities at which these units must be operated to be feasible, substantially all of the electrical energy supplied to the cell is dissipated in overcoming electrical resistance. Combining Ohm's and Faraday's laws we can write

$$E = \text{kwh}/1{,}000 \text{ gal} = k\frac{iARfcF}{ne} \qquad (9)$$

where i = current density, A = transfer area per cell pair, R = total resistance of a unit, f = fraction of electrolyte transferred out of a diluting cell, c = concentration of electrolyte in the diluting cell, F is the faraday constant, n = number of cell pairs in a unit, e = current efficiency, and k is a constant whose value depends on the units of measurement.

Equation (9) shows that the electrical energy required is directly proportional to current density and to salt concentration. Therefore, the electrical energy requirement can be significantly reduced by operating at lower current densities, but the capacity of the unit is directly proportional to the current used and so a balance must be struck between cost of energy and fixed charges on the investment. Equation (9) also shows that E is directly proportional to salt concentration and explains why the energy requirement for sea water is so much greater than for brackish water. In addition to the electrical energy requirement, there is a substantial energy requirement for fluid pumping. For brackish waters of concentration less than 10,000 ppm, the energy requirement in actual installations is of the order of 10 to 30 kwh per 1,000 gallons, which may be less than for the distillation processes. (Very few data are available for distillation processes on brackish waters. Whereas the ideal energy requirement would be much less for brackish waters than for sea water, it is likely that the practical requirement would be substantially independent of initial concentration for a given per cent recovery, but with brackish water the recovery could probably be higher and this would reduce the energy requirement per unit of product.) For sea water, the energy requirement is of the order of 125 to 150 kwh per 1,000 gallons—much higher than for the vaporization or the freezing processes. As is the case for these latter processes, any attempt to reduce the energy requirement by reducing, for example, the current density results in a lower capacity and hence increased fixed costs to obtain the same capacity.

From the extensive data now available from actual operating plants, of which there are more than 20 throughout the world, ranging in size from small 2,000 gpd units up to 2,400,000 gpd, it appears possible to produce fresh water from brackish waters in large plants at a cost within the range of 60 to 100 cents per 1,000 gallons, depending on various local factors. In addition to the energy cost and the fixed costs which comprise 90 to 95 per cent of the total cost in vaporization or freezing processes, membrane replacement is an important item of cost of the electrodialysis process. Another problem in connection with this, or any other process operating away from the seacoast on a brackish water, is how to dispose of the concentrated brine. No general solution to this problem, other than storage in ponds, is in sight at the present time.

Osmionic process. This process, first suggested by Murphy[23], has not progressed beyond the laboratory scale and, in the opinion of the author, is not likely to develop into a process of commercial importance in the foreseeable future. However, it is worth brief discussion because of the interesting scientific principle involved. The process is one of electrodialysis, using the ion-selective membranes, but the actuating energy comes not from an applied emf but from the dilution of a salt solution more concentrated than the one to be separated. The way in which the principle is applied in an actual cell is illustrated by the diagram in Figure 73. At the beginning, compartments (1), (2), and (3) are filled with a brine of concentration C_1 and compartment (0), which surrounds compartments (1), (2), and (3), is filled with concentrated brine at concentration $C_2 > C_1$. Because of the concentration gradient, Cl^- and Na^+ ions from compartment (0) diffuse into compartments (1) and (3), respectively, as shown in Figure 73(b). Maintenance of electrical neutrality in these compartments requires that, simultaneously with

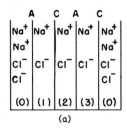

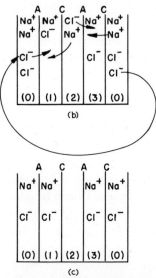

FIG. 73. Principle of the osmionic process.

this process, another one must occur, that of diffusion of Cl⁻ from (2) to (3) and Na⁺ from (2) to (1). The net result is the decrease in concentration of NaCl in compartment (2) as shown in Figure 73(c). The process is seen to require a source of brine of higher salt concentration than the one which is to be demineralized. This could conceivably be obtained by solar evaporation of sea water or from brine wells.

The Dodge-Eshaya Process

This is a vapor-compression, evaporation process based on work by Bliss who conducted laboratory experiments, the results of which are given in a report to the Office of Saline Water [24]. Although the general principles of the process are not new, their embodiment into a working process involved new engineering studies. The process was further developed by Eshaya and Dodge who made an exhaustive engineering and economic analysis of it, the results of which have been reported [11].

The conventional vapor-compression process uses a heating surface submerged in the boiling solution, but in the

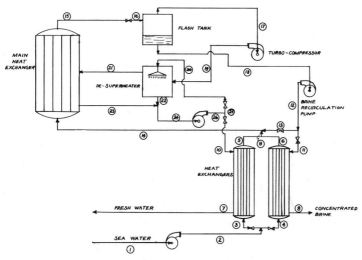

FIG. 74. Diagram of the Dodge-Eshaya process.

Dodge and Eshaya process the solution to be evaporated is pumped at relatively high velocity through the tubes of a heat exchanger as in Figure 74, and the pressure is maintained slightly above the boiling point so that no vaporization occurs. After leaving the exchanger, the solution goes to an expansion valve followed by a flash tank where the solution partially vaporizes, or flashes, as a result of the lowered pressure. The purpose of this combination of nonboiling heat transfer followed by flash evaporation is twofold—to increase the rate of heat transfer and thereby reduce the surface required, and to prevent or at least minimize scale formation. The vapor evaporated in the flash tank goes to a compressor where its pressure is increased slightly (a pressure ratio of about 1.05–1.10 is sufficient) and after desuperheating in a direct spray of some of the condensate, it becomes the heating medium in the main exchanger. Two other special features of the process may be mentioned at this point. One is that the mean temperature difference across the main exchanger is kept very low, only 4 to 5°F, in order to minimize the energy required for vapor compression. The work of compression may be calculated from the steam tables or more simply from the equation

$$\text{horsepower} = \frac{P_1 V_1 (r^2 - 1)}{66{,}000 r \epsilon} \qquad (10)$$

which is a simplified approximation of the usual equation for isentropic compression of an ideal gas that is quite accurate at low pressure ratios [25] where P_1 = inlet pressure in pounds per square foot, V_1 = inlet volume in cubic feet per minute, r = compression ratio, and ϵ = compressor efficiency. Some values for the work in kwh per 1,000 gallons as a function of temperature difference are plotted in Figure 75, and it can be seen that the work requirement increases linearly with Δt. As Δt is reduced, the area required for the transfer of a given amount of heat is increased in accordance with the well known relation

$$q = UA\Delta t \qquad (11)$$

where q is rate of heat transfer in Btu per hour, U = over-all coefficient of heat transfer, Btu per (hour–square foot–°F), A is the transfer surface in square feet, and Δt is the mean temperature difference between the two fluids exchanging heat. To counteract this effect one can try to increase the value of U. The reciprocal of U is an over-all resistance to heat flow which is made up of three resistances in series as expressed in the relation

$$\frac{1}{U} = \frac{1}{h_L} + \frac{L}{k} + \frac{1}{h_S} \qquad (12)$$

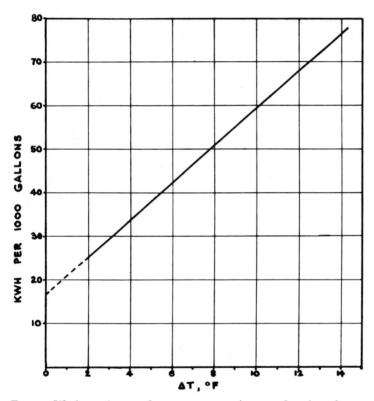

FIG. 75. Work requirement for vapor compression as a function of temperature difference across the main exchanger.

where h_L = coefficient of heat transfer through the liquid film on the brine side of the transfer surface, L/k = resistance of the exchanger wall which is generally negligible, and h_S = coefficient of heat transfer for the condensing steam.

This equation is a simplified one which neglects any differences in area in the flow path and any resistance due to fouling of the surface by scale or other deposits.

Coefficient h_L is a function of fluid velocity of the form

$$h_L = \alpha u^{0.8} \tag{13}$$

where α is a constant whose value depends on the fluid properties and to some extent on the geometry of the system and u is the velocity. Hence, by increasing the rate of flow

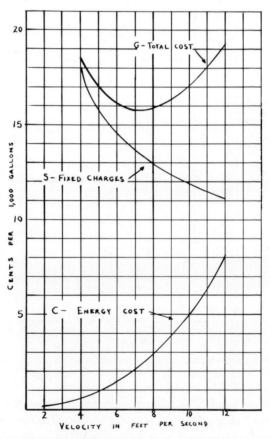

Fig. 76. Economic balance in the determination of the optimum velocity in the main heat exchanger.

through the tubes of the exchanger, a significant increase in h_L and therefore in U can be achieved. At the same time, however, the pumping power is also increased and there is an economic balance between the increasing power cost and the decreasing fixed charges. A number of economic balances of this sort has been made and Figure 76 illustrates one of them [11]. The optimum velocity in the tubes is the one that yields the minimum total cost and, for the specific set of

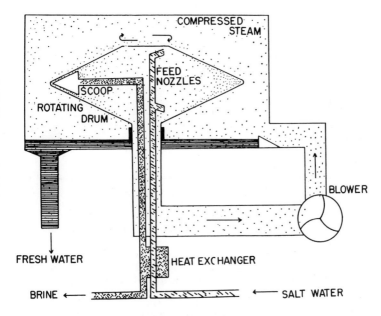

Fig. 77. Diagram of the Hickman rotary still (Courtesy of Dr. K. Hickman and Yale Scientific Magazine, p. 24, May 1959).

conditions in question, the optimum is seen to be about 9 feet per second.

Another approach to the problem of high heat-transfer coefficients on the liquid side is the use of a drum rotating at high speed on whose surface the brine to be evaporated is spread in a thin film by centrifugal force. Figure 77 is a diagrammatic representation of the Hickman rotary still which was developed under contract with the Office of Saline Water and was tested at Wrightsville Beach, North Carolina, in a 25,000 gpd unit. This was a vapor-compression unit as indicated in the diagram, where the so-called "blower" is simply the steam compressor. With this unit, over-all heat-transfer coefficients, U, of the order of 2,000 to 3,000 were reported. The lower of these figures is about the same as would be obtained in the more conventional heat exchanger with a brine velocity of about 10 feet per second in clean one-inch tubes and a steam-film coefficient of 8,000, which is obtainable under conditions of dropwise condensation. This value of U may be compared with a value of about 600 which is generally obtained in the common submerged-tube type of evaporator. Since the heat-transfer area required is inversely proportional to U, it is evident that the forced circulation type of evaporator, including the rotating-surface type, should have four to five times the capacity for the same transfer area.

In my opinion a tubular evaporator surface plus a pump to circulate the liquid at a good velocity is a simpler, less expensive and just as effective device for obtaining high rates of heat transfer as a rotating surface and much easier to scale up to large capacities.

The usual value of the steam-film coefficient, h_S, is 1,500 to 2,000 in Btu per (hour–square foot–°F) and, since h_L has been increased to about 4,000 to 5,000 by the use of good velocities, the steam film has become the limiting resistance to the transfer of heat. The question naturally arises: what can be done to increase h_S? This leads us to a brief discussion of the two types of steam condensation—filmwise and dropwise. On a clean surface that is readily wetted by water, the condensate forms, on the condensing surface, a continuous film of varying thickness depending on various factors. This tends to "blanket" the surface with a nonturbulent layer of water that reduces the rate of heat transfer. On a surface not wetted by water the condensate forms in discrete droplets which quickly run off and continually expose fresh metal surface to the steam. The difference between film and droplet type condensation is illustrated in Figure 78. Often one

(a)

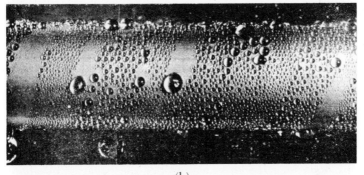

(b)

(c)

FIG. 78. Visual evidence of the two types of steam condensation: (a) film; (b) droplet; (c) mixed [26].

obtains a mixture of the two as shown in the lowest photograph.

With good dropwise condensation, heat-transfer coefficients as high as 20,000 have been reported, but values in the range of 10,000 to 15,000 are common and fairly easy to obtain. One can readily calculate by equation (12) that the over-all coefficient U obtainable with a brine-side coefficient of 3,200, a 16-gauge admiralty-metal tube ($k = 65$) and a steam-side coefficient of 10,000, equals 2,020 compared with a U of 950 for a steam coefficient of 1,500 which is about the usual value. In other words, a more than twofold gain in capacity is possible merely by changing the type of steam condensation from film to droplet.

Under normal conditions steam condenses filmwise, and special means must be used to make it dropwise. This involves coating the metal surface with a substance not wetted by water. There are essentially two ways of doing this, one mechanical and the other chemical. Typical of the former is the placing of a thin film of a material such as Teflon (tetrafluorethane) over the metal surface. This will do the trick but it places another thermal resistance in the path and tends to defeat its own purpose. One can readily calculate that a film of Teflon only 0.00085 inch thick has a thermal resistance equivalent to a steam-film coefficient of 2,000, or that a film of 0.0005 inch thickness will reduce the capacity by about 30 per cent over what it would be if no such film were present and the respective brine and steam coefficients were 2,000 and 10,000.

The chemical method, which seems the more promising, is to treat the metal surface with a "promoter." This is a substance made up of relatively long-chain molecules with a dual nature. One end of the chain consists of group such as carboxyl (—COOH) or a sulfur-containing a group (—SH or —OCSS) which readily reacts with a copper surface and attaches itself firmly as a monolayer. The other end of the chain is a hydrophobic, or water-repellant, type of grouping such as a straight-chain hydrocarbon group of 12 or more carbon atoms. Two typical compounds from among the many that can be used are the following:

$C_{18}H_{37}SH$ (octadecane—thiol)
$C_{17}H_{33}COOH$ (oleic acid)

Some of these compounds are also effective when added to the steam in a proportion of only about 0.03 ppm by weight. This is too small an amount to have any significant effect either on the quality of the water or the cost of treatment.

A considerable amount of small-scale laboratory work on single condenser tubes has been done with dropwise promoters [26–29] and some tests made on commercial condensers with fair success in some cases and disappointing results in others [30–32]. Because of the many factors involved in an extrapolation of the results under controlled conditions in the laboratory on a single tube to a large condenser with many tubes in parallel, much research and development work remains to be done before dropwise condensation can be produced and maintained for long periods in commercial equipment. Considering the very widespread use of steam-condensing equipment and the large potential gains that should accrue from the use of dropwise condensation, a considerable research effort seems justified.

The Scale Problem

Saline waters contain calcium, magnesium, bicarbonate, and sulfate ions as well as sodium and chlorine and, under the operating conditions of many of the processes, precipitates can and will form as a hard scale on the surfaces. The evaporation processes, particularly, have been plagued by scale deposits which are good thermal insulations and greatly reduce the capacity of the equipment. The chief constituents of scale are calcium and magnesium carbonates, magnesium hydroxide, and calcium sulfate. The ions CO_3^{--} and OH^- are not present to an appreciable extent in the saline water but may be formed from the reactions

$$2HCO_3^- = CO_3^{--} + H_2O + CO_2 \tag{14}$$

$$CO_3^{--} + H_2O = CO_2 + 2OH^- \tag{15}$$

The rate of formation will depend on the temperature and the partial pressure of carbon dioxide. Until recently it was necessary to shut down a plant periodically and laboriously remove the scale by hand. Scientific study of the conditions of scale formation have now made it possible to prevent the formation of scale over long periods of time. The most important factors in scale prevention are temperature, degree of concentration of the brine, and hydrogen-ion concentration (pH). All of the scale-forming compounds have inverted solubility curves, in that they become less soluble as the temperature increases. If the temperature is kept below 160°F in an evaporation process, very little scale will form. Carbonates and hydroxides are rendered more soluble by lowering the pH, and scales of these compounds are effectively prevented by maintaining the pH at a value a little below the neutral point, say at about 6, by feeding acids such as sulfuric or salts like ferric chloride. The only means of preventing $CaSO_4$ scale is to avoid exceeding its solubility product, and this limits the allowable concentration of sea water to about twice the initial value, the value depending on other conditions. References are suggested for further information on scale [33-35].

Conclusions

The development of processes for the recovery of fresh water from saline waters is proceeding on a wide front throughout the world. The problem is one of engineering research with the goal of reducing the cost of the product.

The only processes now in large-scale operation on sea water are distillation processes using conventional fuels as the heat source, and the associated cost of the fresh water is of the order of $1.50 to $2.00 per 1,000 gallons. For brackish waters up to, say, 8,000 parts per million of dissolved salts, the electrodialysis process is now in commercial use on a large scale, but insufficient operating experience is available to make reliable cost estimates. This process is not suitable at the present time for the treatment of sea water. Further development is likely to change this and it has been estimated that in large plants, water could be produced from sea water at a cost of 50 to 100 cents per 1,000 gallons and at a much lower cost from brackish waters.

The two main items of cost in most of the processes are energy and fixed charges on the investment. The thermodynamic minimum energy requirement is about 3 kilowatt hours per 1,000 gallons of pure water for zero per cent recovery and about 4 kwh for 50 per cent recovery. The energy cost would be a minor one if these figures could be approached in practice, but actual processes have thermodynamic efficiencies of only 2 to 6 per cent. Increasing the efficiency is one of the main goals of present research in the field but this is not so easy as might appear to a non-engineer. Any attempt to decrease energy requirement means reducing the irreversible effects and hence the driving forces, and this inevitably leads to larger and more costly equipment. Therefore, regardless of the process considered there is always an economic balance between the decreasing cost of energy and the rising costs based on the investment. It appears that the optimum point of minimum cost will occur at around 10 to 15 per cent efficiency for the best processes, assuming 50 per cent recovery from sea water. In other words, the practical energy requirement is not likely to be less than 30–40 kwh per 1,000 gallons.

The distillation process, in one of its several forms, is likely to remain the most important one for large-scale use on sea water for some years. The freezing process, also with several variants, is now emerging from the laboratory stage and may very well compete with distillation in a few years. Many other processes have been proposed, including sources of energy other than conventional fuels, such as solar energy and nuclear energy, and are being actively investigated. None of them at this stage appears likely to become a serious competitor for distillation or freezing processes. Some of these processes will probably find limited application under special conditions.

Finally, it seems reasonable to predict that fresh water will be made in the near future from sea water in large plants at costs ranging from 50 to 100 cents per 1,000 gallons at the seacoast. This is still appreciably higher than the average delivered cost of municipal water in the United States and far above the cost of industrial and irrigation water. Nevertheless, it is comforting to know that we have an inexhaustible supply of water to which we can turn when the necessity arises and, whereas the cost will be greater than we are used to paying, it will not be prohibitive for the majority of uses. With brackish water as the source, appreciably lower costs per 1,000 gallons may be expected.

With some of the distillation processes the production of electric power as a by-product may materially reduce the cost of the water. Chemical by-products from the desalination of sea water are not to be expected as none of the processes carries the concentration far enough to yield any solid products of value.

REFERENCES

1. *A Standardized Procedure for Estimating Costs of Saline Water Conversion.* Office of Saline Water, U.S. Dept. of the Interior (March 1956).
2. GILLILAND, E. R. *Ind. Eng. Chem.*, **47,** 2410 (1955).
3. DODGE, B. F., and A. M. ESHAYA. *Advances in Chem.* Ser. **27** (1960).

4. *Study of the Applicability of Combining Nuclear Reactors with Saline Water Distillation Processes*, PB 161062. Office of Saline Water, U.S. Dept. of the Interior.
5. ELLIS, C. *Fresh Water from the Ocean*. New York, Ronald Press (1954).
6. *Proceedings, Symposium on Saline Water Conversion*. Res. Council, Nat. Acad. Sci. (U.S.), Publ. 568, pp. 117–176 (Nov. 1958).
7. LÖF, G. O. *Demineralization of Saline Water with Solar Energy*. Report No. 4. Office of Saline Water, U.S. Dept. of the Interior (1954).
8. TELKER, M. *New and Improved Methods for Lower Cost Solar Distillation.* Report No. 31. U.S. Dept. of the Interior, Office of Saline Water (1959).
9. *Demineralization of Saline Waters*. A Preliminary Discussion of a Research Program with an Outline and Description of Potential Processes and a Bibliography. U.S. Dept. of the Interior (Oct. 1952).
10. BADGER, W. L., and F. C. STANDIFORD. See Ref. 6; pp. 103–114.
11. DODGE, B. F., and A. M. ESHAYA. *Economic Evaluation Study of Distillation of Saline Water by Means of Forced-circulation Vapor-compression Distillation Equipment*. Report No. 21. Office of Saline Water, U.S. Dept. of the Interior.
12. ANTHONY, P., and L. BERKOWITZ. See Ref. 6; pp. 80–90.
13. BOSWORTH, C. M., S. S. CARFAGNO, A. J. BARDULIN, and D. J. SANDELL. *Further Development of a Direct-freezing Continuous Wash-separation Process for Saline Water Conversion*. Report No. 32. Syracuse, N.Y., Carrier Corp. (July 1959).
14. PARENT, J. D. *The Storage of Natural Gas as Hydrate*. Bull. No. 1., Chicago, Institute of Gas Technology (1948).
15. HIMES, R. C., S. E. MILLER, W. H. MINK, and H. L. GOERING. *Ind. Eng. Chem.*, **51,** 1345 (1959).
16. DAVISON, R. R., and D. W. HOOD. See Ref. 6; pp. 400–416.
17. SOLLNER, K. *J. Phys. Chem.*, **49,** 47 (1945).
18. GREGOR, H. P. See Ref. 6; p. 240.
19. MURPHY, G. W., and R. C. TABER. See Ref. 6; pp. 196–211.
20. JUDA, W., T. A. KIRKHAM, and E. A. MASON. See Ref. 6; pp. 265–282.
21. VOLCKMAN, O. B., and W. H. MOYERS. See Ref. 6; pp. 283–315.
22. MENTS, M. V. *Ind. Eng. Chem.*, **52,** 149 (1960).
23. MURPHY, G. W. *Ind. Eng. Chem.*, **50,** 1181 (1958).
24. BLISS, H. Forced circulation and dropwise condensation techniques for improving heat transfer rates for vapor compression evaporators. Report No. 8. Office of Saline Water, U.S. Dept. of the Interior (Oct. 1955).

25. DODGE, B. F. *Chemical Engineering Thermodynamics*, p. 355. New York, McGraw-Hill (1944).
26. BLACKMAN, L. C. F. *Research*, **11**, 394 (1958).
27. BLACKMAN, L. C. F., and M. J. S. DEWAR. *J. Chem. Soc.*, Jan., p. 171 (1957).
28. BLACKMAN, L. C. F., M. J. S. DEWAR, and H. H. HAMPSON. *J. Appl. Chem.*, **7**, 160 (1957).
29. DREW, T. B., W. M. MOYLE, and W. Q. SMITH. *Trans. Am. Inst. Chem. Engrs.*, **31**, 605 (1935).
30. BRUNT, J. J., and J. W. MINKEN. *Ind. Chemist*, May, p. 219 (1958).
31. GARRETT, D. F. *Brit. Chem. Eng.*, **3**, 498 (1958).
32. BIRT, D. C. P., J. J. BRUNT, J. T. SHELTON, and R. G. H. WATSON. *Trans. Inst. Chem. Engrs. (London)*, **37**, 289 (1959).
33. *Critical Review of Literature on Formation and Prevention of Scale*. Report No. 25. Office of Saline Water, U.S. Dept. of the Interior (July 1959).
34. HILLIER, H. *Proc. Inst. Mech. Engrs. (London)*, **IB**, 295 (1952).
35. LANGELIER, W. F., D. H. CALDWELL, and W. B. LAWRENCE. *Ind. Eng. Chem.*, **42**, 126 (1950).

INDEX

Adiabatic gradient, 56, 61, 63–65, 67, 69–70
Alkaloids: alstonidine, 209–12; alstoniline, 207–08; alstonine, 199–207; biosynthesis, 117 ff., 124–37, 141–42; definition of, 196; of *Alstonia muelleriana*, 212–15; plant, 118 ff.; sempervirine, 202–03; tobacco, 119 ff.
Alstonia, alkaloidal constituents of, 198–99, 207. *See also* Alstonine
Alstonia muelleriana: extraction of, 212–14; hypotensive effect of, 214–15
Alstonidine, 209; spectrum of, 210–12
Alstoniline, 207–08
Alstonine: antimalarial properties of, 199; molecule, 199–204; relation to serpentine, 204–06; spectrum of, 203–05
Alstyrine, 200, 202
Anabasine, 119–20, 123–24, 131–33, 135–36
Analogy, 99–100, 105–06
Animals, sex reversal in. *See* Sex
Arts: communication, 93–96; criteria for, 111–16; definition of, 98–101; disciplines of, 101–06; imitation of science in, 113–15; importance to scientists, 93–94, 109–10; similarity to science, 107–10
Aruba, water conversion plant at, 280–81, 288–89
Australian trees, 195 ff.
Autoradiography, 152–56, 164–66

Basalt: glass, 13–15, 48; lava, 44, 68–69
Biosynthesis of alkaloids, 117 ff., 124–37, 141–42
Black-white discrimination (rhesus monkeys), 248–50
Blastophthoria, 184–86, 189
Blood pressure, hypotensive agents for, 195–97, 202, 206–07, 214–15
Brackish water, 273–74, 277, 306, 310–11, 323–24

Callus tissues, 137–39, 218–19, 221, 224–25, 234–36
Carboxyl displacement, 125–31
Cell: differentiation, 167–68; division, 145, 153, 166, 219–20, 223–26, 228, 235–37
Cells: fuel, 281, 283; germ, 171, 174–77, 181, 185–86, 226–28
Chromatography, 81–85, 212–14
Chromosome: aberrations, 189–91; conjugation, 172–74; composition of, 145–52; duplication, 152–59, 163, 166; exchange, 155–63; reproduction, 145, 151–52, 155; sex, 176, 179–82, 188; subunits, 155–63
Claude process, 282
Coconut milk, 219, 223–24
Codehydrogenases, 129
Codes: genetic, 150–51, 163–68, protein, 147 ff., 150–51
Colorado plateau, 5–6
Continents: age and erosion of, 11–12, 40–41; compared with ocean ba-

sins, 11; composition of, 3–4, 43–44, 48–49, 51; elevation of, 40; heat flow, 51–52; plateaus of, 4–7, 12, 17; sialic, 3–12, 16–17, 19; structure, 1–3, 13, 15–20; troughs of, 7–8, 12, 18. *See also* Earth

Convection: currents, 6–7, 17; in core, 69–71; in mantle, 55–56, 68–69. *See also* Adiabatic gradient

Counter current extraction. *See* Chromatography

Critical pressure process, 284, 294–96

Crystallization or freezing process, 284–85, 296–302

Cybernetics, 85

Demineralization of saline water, 281–324. *See also* Water

Deoxyribonucleic acid (DNA), 146–47; molecule, 147–53, 156–57, 162–67

Discrimination (rhesus monkeys): black–white, 248–50; object, 249–54, 257–61; spatial, 246–48, 251

Distillation process. *See* Evaporation

DNA. *See* deoxyribonucleic acid

Dodge–Eshaya process, 313–17

Drosophila, 179–80

Earth: crust, 1–20, 41–44, 51, 65; density, 43, 64–65; heat flow in, 8–12, 16, 19, 41–42, 49–56; inner core, 43–44, 46, 49, 59–60, 62, 64–69, 71–72; inner-outer core boundary, 58, 62–63; magnetic field of, 69–72; mantle, 43–44, 46, 48, 52–55, 59–60, 62, 64–69; mantle–core boundary, 58, 60–62, 66; outer core, 3, 43–44, 46–47, 49, 60, 62, 67, 69–71; pressure in, 45–46; pressure–temperature gradient in, 15–18; radiogenic heat of, 9, 47–49, 72–73; structure of, 1–3, 5–20, 42–47; temperature, 39–74; thermal history of, 41, 47–50, 52, 54–55, 72–74

Earthquake waves, 3, 42–43, 45

Eclogite, 13–14, 19

Einstein, Albert, 25–26, 96, 110

Electrodialysis process, 273, 284, 306–11, 323

Endosperm, 222–24

Energy: requirements and costs of, 275–80, 286, 323; sources for demineralization, 280–83 *(see also* Water)

Eridanus constellation, 25

Erosion, 4–5, 11–12, 17–18, 40–41

Evaporation or distillation processes, 284–96, 324

Extraction of *Alstonia muelleriana*. *See* Chromatography

Extraction, counter current. *See* Chromatography

Feldspar, 14–15

Ferns, 119, 218, 225–32

Fertilization, 145–46, 172, 183–84; delayed, 184–86, 189; of ferns, 227–28

Freezing process. *See* Crystallization

Gabbro, 14–15, 19, 48

Genetic heritage: coding of, 145 ff., 150–51, 163–68; transmission of, 145 *(see also* Chromosomes)

Genetic sex determination, 176–80

Germ: cells, 174–77, 181, 185; plasm, 174–77

Gonadal agenesis: complete, 186–88; partial, 188–89

Gonochorists, 179

Grand Canyon, 5

Gravitation, 1, 18, 24–25, 31, 64–65, 95–98, 109–10

Growth rate, correlation with nicotine production, 122–24, 134–37, 141–42

Guessing, 81–84, 89

Gulf of Mexico, 7, 11

Hamilton perseverance test, 262–64

Harman, 200, 210

Heat: conduction equation, 49–51,

INDEX

54–57, 70–71; dissipation, in earth, 39–41; exchanger, 313–17; flow, 8–12, 16, 19, 41–42, 49–56; radiogenic, 39, 47–49, 72–73; transfer, 313–21

Hermaphrodites, 177–79, 183; in man, 189–90

Hertzsprung–Russell diagram, 23, 32, 35

Hickman rotary still, 317–18

Honesty, scientific. See Scientific

Humanities, 93 ff. See also Arts

Hydrocarbon hydrates process, 284, 301–02

Hydrogen bonds, 148–50, 167

Ideas, scientific selection of, 83–86

Infant learning, 240–42, 250–52, 266–67. See also Learning

Inner core–outer core boundary. See Outer core

Intellectual curiosity, 78

Inversion, polymorphic, 13

Isostasy, 4–7

Isotopes, 9, 119, 155

Learning: conditioned response, 242–45, 251; delayed response, 254–57; development in rhesus monkeys, 239 ff.; object discrimination, 249–54, 257–61; of children, problems in studying, 240–41; spatial discrimination, 246–48, 251; straight-runway performance, 245–46, 251

Magnetic field, of earth. See Earth

Man: sex deviations in, 185–91; sex reversal in animals and, 171 ff.

Mantle. See Earth

Maze solution, 246–47, 251

Melting point, of rocks. See Rocks

Mohorovičić (M, Moho) discontinuity, 3–5, 14–20, 43–44, 65, 67

Monkeys, rhesus: learning development in, 239 ff.; in first 60 days, 242, 267–68; in first year, 250–61; terminal learning ability, 261–68. See also Learning

Morphogenesis: early, 219–26, 235–37; effect of environment on, 231–32, 236–37; in ferns, 226–32; later phases of, 232–37; processes, 136–38

Moths, temperature influence on, 180

Motives for scientific inquiry. See Scientific

Mountains: age and erosion of, 5, 11–13, 17–18, 40–41; formation of, 8, 40–41; heat flow in, 5, 9, 15–16; origin of, 1 ff., 18; structure, 1–4; temperatures in, 52. See also Earth

Mutation: effects on man, 176–77, 185

Newton, Isaac, 80, 85, 89, 96–98, 109–10

Nicotiana: alkaloids, 117 ff., 119; biosynthesis and growth of, 124–37, 141–42; molecular structure, 120, 124–31; root system, 121–24, 131–37; species, 139–41

Nicotine, 119–20, 131–42

Norharman, 201

Nornicotine, 120

Nucleic acid, 146–47, 167. See also Deoxyribonucleic acid, Ribonucleic acid

Nucleotides, 127–28, 147–51, 163

Object discrimination. See Discrimination

Oceans: basins, 1–3, 7, 11; crust, 43–44; floor, heat of, 9–11, 15–17, 19–20, 40, 51–55, 65; rocks, 3, 8, 12, 51

Oddity test, 264–67

Office of Saline Water (OSW), 273, 284, 290, 297, 299, 313, 318

Ornithine, 125–26, 130, 132–33, 141

Osmionic process, 311–12

Outer core, 3, 43–44, 46–47, 49, 60, 62, 67; –inner core boundary, 58, 62–63

Pacific ocean, heat flow in, 52–55. See also Oceans
Phenylalanine, 130
Philosophy, 102–06, 108, 114; analytical, 114
Plant alkaloids. See Alkaloids
Plants, morphogenesis in, 217 ff. (see also Morphogenesis)
Polymorphic inversion, 13
Pressure: distribution in the earth, 45–46; high, effect on earth, 13–15, 17–18, 57–58; relation to melting points, 60–63, 66; –temperature gradient, 15–18, 50–51
Proteins, 147 ff.; codes, 150–51 (see also Codes, genetic)

Radiation laws, 24
Radioactivity: in earth, 9–10, 16, 49–51, 54–56, 68, 72–73; in cells, 152
Radiogenic heat, 39, 47–49, 72–73
Rauwolfia serpentina, 202
Refrigeration process, 284, 296–300
Relativity Theory, 25–26, 96, 110
Reproduction: of chromosomes, 145 ff. (see also Chromosomes); vegetative, 217 ff. (see also Morphogenesis)
Reserpine, 202, 214–15
Reverse-osmosis process, 284, 304–05
Ribonucleic acid (RNA), 146, 164–67
Rocks: convection currents in, 6, 17, 55–56; effect of high pressure on, 13–15; heat generation in, 6, 8–15, 47–54; in mantle, 3; melting point of, 58–63; sialic, 3–12, 16–17, 19; simatic, 2–5, 8–11; types of, 2–5, 17–19, 43–44, 50–51
Root: cultures, 121–32, 137–38, 219–21; system (tobacco), 121–24, 131–37

Saline Water, Office of. See Office of Saline Water
Salt, separation from saline water, 305 ff. See also Water

Scale formation, 321–22
Science: applied, 77–81, 87, 271–72; basic, 77–81, 87, 196–97, 271–72; compared with arts, 115–16 (see also Arts); cultivation of, 77 ff.; definition of, 101; literary imitations of, 113–14; specialization in, 86–87
Scientific: achievement, evaluation of, 77, 85, 87–89, 111–15; communication, 93–100, 107–11; curiosity, 78–80; formulae, 96–100; honesty, 89–90; inquiry, motives for, 77–81, 84–85; method, 77, 81–87, 110–11
Sediments, 7–8, 11–12, 18; radioactive, 9, 18
Serpentine, 202–07
Sex: chromosomes, 176, 179–82, 188; determination, 176–80; deviations in man, 185–91; reversal in animals and man, 171 ff.; reversal, postgenetic, 181–84
Sirius, 23–24; companion, 24–25 (see also White dwarfs)
Solar evaporation process, 281–84, 293–94
Solvent extraction process, 284, 302–03
Spatial discrimination, 246–48, 251
Spectra: of alstonidine, 210–12; of alstonine, 203–05; of harman, 210; of sempervirine, 202–03; of serpentine, 203–05; of white dwarfs, 24–30, 36
Sperm: chemical contents, 146; function, 145–46; mutation changes in, 175–77
Stars: degenerate, 25–26, 34–37; dwarf, 23–37; evolution of, 23 ff., 32–36; spectra of, 24–30, 36; Zwicky-Humason, 28, 36
Supercritical pressure process, 284, 294–96

Temperature: distribution in earth, 49–58, 64–65, 67–68; gradients of

INDEX

earth, 9, 16, 50, 56; influence on moths, 180; of earth, 39 ff.; of mantle, 52–55, 58, 61–65, 67–69; of white dwarfs, 24–26, 29, 32, 36. See also Heat

Thermal: expansion, 65–67; history of earth, 41, 47, 50, 52, 54–55, 72–74

Thymidine, 152. See also Tritium-labeled

Tibetan plateau, 6

Tobacco alkaloids. See Nicotiana

Tritium-labeled: cytidine, 164–66; thymidine, 153–55

Universality: loss of, 86–87; of arts and sciences, 107

University, structure of, 102–06

Vapor-compression evaporation process, 284, 291–93

Vegetative reproduction, 217 ff. See also Morphogenesis

Volcanoes, 8, 39–40, 54, 68

Water: basic facts about, 272–74; brackish, 273–74, 277, 306, 310–11, 323–24; conversion plant at Aruba, 280–81, 288–89; cost of conversion of saline to fresh, 274–84, 323–24; processes for conversion of saline to fresh, 280–324

White dwarfs: age of, 32–34; definition of, 35–37; density of, 30–31; discovery of, 23–29; mass of, 29–30; observation of, 29; spectra of, 24–30, 36; temperature of, 24–26, 29, 32, 36

Xenopus, 182–84

Zwicky–Humason stars. See Stars